RIGOROUS MATHEMATICAL THINKING

This book demonstrates how rigorous mathematical thinking can be fostered through the development of students' cognitive tools and operations. Though this approach can be applied in any classroom, it seems to be particularly effective with socially disadvantaged and culturally different students. The authors argue that children's cognitive functions cannot be viewed as following a natural maturational path: They should be actively constructed during the educational process. The Rigorous Mathematical Thinking (RMT) model is based on two major theoretical approaches that allow such an active construction of cognitive functions: Vygotsky's theory of psychological tools and Feuerstein's concept of mediated learning experience. The book starts with general cognitive tools that are essential for all types of problem solving and then moves to mathematically specific cognitive tools and methods for utilizing these tools for mathematical conceptual formation. The application of the RMT model in various urban classrooms demonstrates how mathematics education standards can be reached even by students with a history of educational failure who were considered hopeless underachievers.

James T. Kinard, Sr., earned his Ph.D. in electroanalytical chemistry from Howard University and is president of Innovations for the Development of Cognitive Literacy, Inc., Chicago, Illinois. He developed and implemented the Rigorous Mathematical Thinking program and is a certified trainer of the Feuerstein cognitive development program, Instrumental Enrichment. He lectures at international cognitive enrichment workshops in the United States, Canada, the United Kingdom, France, The Netherlands, and India.

Alex Kozulin is research director of the International Center for the Enhancement of Learning Potential in Jerusalem, Israel, and teaches at Tel Aviv University and Hebrew University. He held an academic appointment at Boston University, was a visiting professor at the University of Exeter and at the University of Witwatersrand, and was a visiting scholar at Harvard University. Dr. Kozulin is author of *Vygotsky's Psychology: A Biography of Ideas* (1990) and *Psychological Tools: A Sociocultural Approach to Education* (1998).

Rigorous Mathematical Thinking

Conceptual Formation in the Mathematics Classroom

JAMES T. KINARD, Sr.
Innovations for the Development of Cognitive Literacy

ALEX KOZULIN
International Center for the Enhancement of Learning Potential

Shaftesbury Road, Cambridge CB2 8EA, United Kingdom

One Liberty Plaza, 20th Floor, New York, NY 10006, USA

477 Williamstown Road, Port Melbourne, VIC 3207, Australia

314–321, 3rd Floor, Plot 3, Splendor Forum, Jasola District Centre, New Delhi – 110025, India

103 Penang Road, #05–06/07, Visioncrest Commercial, Singapore 238467

Cambridge University Press is part of Cambridge University Press & Assessment, a department of the University of Cambridge.

We share the University's mission to contribute to society through the pursuit of education, learning and research at the highest international levels of excellence.

www.cambridge.org
Information on this title: www.cambridge.org/9780521876858

First published 2008

A catalogue record for this publication is available from the British Library

Library of Congress Cataloging-in-Publication data
Kinard, James T., 1943–
Rigorous mathematical thinking : conceptual formation in the mathematics classroom / James T. Kinard, Sr., Alex Kozulin.
p. cm.
Includes bibliographical references and index.
ISBN 978-0-521-87685-8 (hardback) – ISBN 978-0-521-70026-9 (pbk.)
1. Mathematics – Study and teaching. 2. Mathematics – Psychological aspects.
I. Kozulin, Alex. II. Title
QA11.2.K56 2008
510.71 – dc22 2007050177

ISBN 978-0-521-87685-8 Hardback
ISBN 978-0-521-70026-9 Paperback

Contents

Introduction

"There are 26 sheep and 10 goats on a ship. How old is the ship's captain?" This and similar tasks were given during the math lessons to primary school students in a number of European countries. More than 60% of students attempted to solve the problem by combining the given numbers, for example, by adding the number of sheep and the number of goats (Verschaffel, 1999). In our opinion, students' handling of the "Captain" problem is emblematic of the difficulties experienced by many students in the math classrooms because it clearly demonstrates that the students' main difficulty was not with mathematical knowledge but with more general cognitive functions that form prerequisites of mathematical reasoning. Students who blindly started to apply mathematical operations to the numbers given in the task ignored a host of cognitive operations that are needed for any sensible problem solving. They neither oriented themselves in the given data, nor compared or classified it. They also did not formulate the problem presented in this task, most probably because no one taught them the difference between the question ("How old . . . ") and the task's real problem. They apparently were not used to thinking of the tasks as having one solution or several or an unlimited number of correct solutions or no solution at all. For them, mathematics apparently appeared as an associative game where the winner correctly guesses which standard operation fits which one of the standard tasks.

In this book we attempt to demonstrate how rigorous mathematical thinking can be fostered through the development of cognitive tools and operations. Though our approach can be applied in any classroom, it seems to be particularly effective with socially disadvantaged and culturally different students. We will start with more general cognitive tools that are essential for all types of problem solving and then move to mathematically specific cognitive tools. Such an approach is based on our belief that although mathematics, as we know it today, represents an integration of elements from a number of cultures, it has its own unique culture that is distinctively different from "everyday ways" of doing things in various societies and cultures. Cognitive functions that appear naturally following the maturational path in one culture immediately reveal their culturally constructed nature once observed in children belonging to a different cultural group. Thus one cannot take for granted a certain type of cognitive development in students of a multicultural classroom. Their cognitive functions, both of a general and a more specifically mathematical nature, should be actively constructed during the educational process. Our rigorous mathematical thinking (RMT) model is based on two major theoretical approaches allowing such an active construction – Vygotsky's (1979; see also Kozulin, 1998a) theory of psychological tools and Feuerstein's (1990) theory of mediated learning experience.

Chapter 1 starts with a description of mathematical culture as having slowly developed over centuries from sociocultural needs systems rather than isolated, spontaneous efforts of individual scientists. A needs system is a set of internalized habits (autonomous ways of doing things), orientations (preferences and perspectives), and predispositions (inclinations and tendencies) that work together to provide the "blueprints" for actions and the meanings for developing know-how. Sociocultural needs systems are integrally and functionally bound to the life and "ways of living" of the human society. Their nature is an intertwining of affective and cognitive dimensions. Among the most prominent of these systems relevant to the mathematics culture are the needs for spatial and temporal orientation, determination of part/whole relationships, evaluation and establishment of constancy and change, order and organization, and so on. We then proceed to define the concepts of mathematical activity and mathematical knowledge.

The goal of mathematical learning is the appropriation of methods, tools, and conceptual principles of mathematical knowledge with efficient cognitive processing constituting an essential prerequisite of mathematical learning. Such a definition is based on the extension of Vygotsky's notion of learning activity (discussed in Chapter 3) to the domain of mathematical classroom learning. To achieve this objective we begin with identifying and elaborating specific criteria for determining which actions in the mathematics classroom meet the RMT standard. All of the following three criteria must be met for any action to qualify as a mathematical learning activity: (a) the action must contribute to creating a structural change in the students' understanding of mathematical knowledge; (b) the action must aim toward, and therefore be a part of, a systemic process for constructing a mathematics concept, because all concepts in mathematics are characterized as "scientific" according to Vygotsky (1986); and (c) the action must introduce the students to the language and rules of mathematics culture with regard to how things are done in mathematics.

Mathematical knowledge consists of organized, abstract systems of logical and precise understandings about patterns and relationships. These patterns and relationships may not originate in the everyday experience of the child, which, however, does not disqualify them as one of the sources for comprehending this experience mathematically. Mathematical knowledge exists at three levels: mathematical procedures and operations, mathematical concepts, and mathematical insights. Mathematical operations involve basic processes of organizing and manipulating mathematical information in meaningful ways that support and build important ideas and concepts. All mathematical concepts are "scientific" according to Vygotsky's (1986) definition of this term, that is, they are theoretical, systemic, and generative. Mathematical insight is derived from one or more of these conceptual understandings, forming relationships between these understandings, and constructing new ideas and/or applications.

In the RMT paradigm specific, well-defined cognitive processes drive mathematical operations and procedures. Mathematically specific cognitive tools, through their structure/function relationships, organize and integrate the use of cognitive processes and mathematical operations to systemically construct mathematical conceptual understandings. This rigorous practice of conceptual formation develops the students' habits of mind and a propensity for mathematical theoretical thinking and metacognition. These qualities position the student to make higher level reflections about patterns and relationships and create mathematical insights.

The next concept to be introduced is that of psychological tools. Mathematically specific psychological tools extend Vygotsky's (1979) notion of general psychological tools. Symbolic devices and schemes that have been developed through sociocultural needs to facilitate mathematical activity that, when internalized, become students' inner mathematical psychological tools. The structuring of these tools has slowly evolved over periods of time through collective, generalized purposes of the transitioning needs of the transforming cultures. Among the most prominent mathematically specific psychological tools are place value systems, number line, table, x-y coordinate plane, equations, and the language of mathematics. The problem in current mathematics instruction is that these devices are perceived by students as pieces of information or content rather than as "tools" or "instruments" to be used to organize and construct mathematical knowledge and understanding. Both the creation of such tools and their utilization develop, solicit, and further elaborate higher order mental processing that characterizes the dynamics of mathematical thinking. In this regard, the language of mathematics serves both as a tool and a higher order mental function. In the RMT paradigm the instrumentality of the language of mathematics can be viewed from the perspective of how it organizes and transforms students' everyday language and spontaneous concepts into more unified, abstract, and symbolic expressions.

Any genuine mathematical reasoning is rigorous. We define mental rigor as that quality of thought that reveals itself when students' critical engagement with material is driven by a strong, persistent, and inflexible desire to know and deeply understand. When this rigor is achieved, the learner is capable of functioning both in the immediate proximity as well as at some distance from the direct experience of the world and has an insight into the learning process, which has been described as metacognitive. This quality of engagement compels intellectual diligence, critical inquiry, and intense searching for truth – addressing the deep need to know and understand. Rigor describes the quality of being relentless in the face of challenge and complexity and having the motivation and self-discipline to persevere through a goal-oriented struggle. Rigorous thinking requires an intensive and aggressive mental engagement that dynamically seeks to create and sustain a higher quality of thought. Rigorous thinking can thus be characterized by sharpness in focus and perception; clarity and completeness in definition; delineation of critical attributes, precision, accuracy; and the depth of comprehension and understanding.

Chapter 2 focuses on the relationship between the RMT paradigm and the goals and objectives of mathematics education. The overarching goal of education in the United States is to prepare students to function as productive citizens in a highly industrialized and technical society. Since the 1960s there have been numerous attempts to reform education so that it provides a greater focus on scientific and mathematical literacy. One of the most recent attempts in this direction has been the standards-based movement, which has developed specific requirements for each learning subject at each grade level. However, these standards were often formulated in terms of the product of education rather than its process. Benchmarks were established that served as both frameworks and guidelines for curricula and anticipated milestones for student achievement. In spite of all of the good intentions of the standards movement, the current approach to teaching science and mathematics concepts in U.S. classrooms involves the presenting and eliciting of ready-made definitions with accompanying activities that, at best, produce little understanding and superficial applications. The focus in the applications usually does not extend beyond the mechanics or algorithms required for producing concrete answers. Students are not rigorously engaged in developing and manipulating the deeper structures of their thinking, nor are they challenged to synthesize from their own experiences and knowledge base the understanding necessary to induce the abstractions and generalizations that underlie science and mathematics concepts. Thus, many students complete courses in science and mathematics with the illusion of competency based on memory regurgitation. They do not build the understanding nor the flexible structures required for genuine transfer of learning and the creation of new knowledge in various contexts and situations. These surface experiences are not meaningful to students, do not promote science and mathematics competencies, and to some extent contribute to higher dropout rates.

To better understand stronger and weaker aspects of the standards movement, it is instructive to look at the difference between the American system and other systems of education. The study of Stigler and Hiebert (1997) demonstrated that U.S. 8th-grade students scored below their peers from 27 nations in mathematics and below their peers from 16 nations in science. The average international level, however, is also far from adequate. These and other research findings point to two gaps in students' mathematics and science academic achievement: overall, U.S. students perform below students from some other nations and students internationally perform well below expectations, particularly with regard to conceptual mastery. A third gap is the performance of minority students versus that of white students in the Untied States. The African American/white and Latino/white academic achievement gaps in mathematics in the United States widened in the 1990s after African American and Latino students' performance improved dramatically during the 1970s and 1980s.

For the standards movement to succeed, three critical needs should be addressed. First and foremost, U.S. students, and indeed all students, must develop the capability and drive to do rigorous higher order mathematical and scientific thinking. Second, high school students must develop a deep understanding of big ideas in mathematics and science and be able to apply them across various disciplines and in everyday living. Third, students must be able to communicate and express their mathematical and scientific thinking orally and in writing with precision and accuracy. It is imperative that the U.S. mathematics and science education enterprise make serious, substantial, and sustained investments in addressing these needs for real academic achievements and transfer of learning to take place for all students.

Chapter 3 demonstrates the relevance of Vygotsky's sociocultural theory for mathematics learning. For a long time, the predominant model of school learning was that of direct acquisition. Children were perceived as "containers" that must be filled with knowledge and skills. In time it became clear that the acquisition model is insufficient both theoretically and empirically. On the one hand, children have proved to be much more than passive recipients of information; on the other hand, students' independent acquisition has often led to the entrenchment of immature concepts and "misconceptions" as well as a neglect of important academic skills. A search for an alternative learning model brought to the fore such concepts as mediation, scaffolding, apprenticeship, and design of learning activities.

Vygotsky's (1986) theory stipulates that the development of the child's higher mental processes depends on the presence of mediating agents in the child's interaction with the environment. Vygotsky himself primarily emphasized symbolic tools-mediators appropriated by children in the context of particular sociocultural activities, the most important of which he considered to be formal education. Russian students of Vygotsky researched two additional types of mediation – mediation through another human being and mediation in a form of organized learning activity. Thus the acquisition model became transformed into a mediation model. Some mediational concepts such as scaffolding or apprenticeship appeared as a result of direct assimilation of Vygotsky's ideas; others like Feuerstein's (1990) mediated learning experience have been developed independently and only later became coordinated with the sociocultural theory.

In Vygotsky's sociocultural theory, cognitive development and learning are operationalized through the notion of psychological tools. Cultural-historical development of humankind created a wide range of higher order symbolic tools, including different signs, symbols, writing, formulae, and graphic organizers. Individual cognitive development and the progress in learning depend, according to Vygotsky, on the students' mastery of symbolic mediators and their appropriation and internalization in the form of inner psychological tools.

Mathematical education finds itself in a more difficult position vis-à-vis symbolic tools than other disciplines. On the one hand, the language of mathematical expressions and operations offers probably the greatest collection of potential psychological tools. On the other hand, because in mathematics everything is based on special symbolic language it is difficult for a student, and often also for a teacher, to distinguish between mathematical content and mathematical tools. One may classify psychological tools into two large groups. The first is general psychological tools that are used in a wide range of situations and in different disciplinary areas. Different forms of coding, lists, tables, plans, and pictures are examples of such general tools. One of the problems with the acquisition of these tools is that the educational system assumes that they are naturally and spontaneously acquired by children in their everyday life. As a result, general symbolic tools, such as tables or diagrams, appear in the context of teaching a particular curricular material and teachers rarely distinguish between difficulties caused by the students' lack of content knowledge and difficulties that originate in the students' poor mastery of symbolic tools themselves. The lack of symbolic tools becomes apparent only in special cases, such as a case of those immigrant students who come to the middle school without prior educational experience. For these students, a table is in no way a natural tool of their thought, because nothing in their previous experience is associated with this artifact.

Another of Vygotsky's concepts relevant to the task of developing RMT is the zone of proximal development (ZPD) – one of the most popular and, at the same time, most poorly understood of Vygotsky's theoretical constructs (see Chaiklin, 2003). From the perspective of math education, the developmental interpretation of ZPD calls for the analysis of those emerging psychological functions that provide the prerequisites of rigorous mathematical reasoning. Several questions can be asked here. For example, the emergence of which psychological functions is essential for successful mathematical reasoning at the child's next developmental period? What type of joint activity is most efficient in revealing and developing these functions in the child's ZPD? What characterizes the students' mathematically relevant ZPD at the primary, middle, and high school periods? These questions are directly related to the issue of the relationship between so-called cognitive education and mathematical education. There are reasons to believe that the students' mathematical failure is often triggered not by the lack of specific mathematical knowledge but by the absence of prerequisite cognitive functions of analysis, planning, and reflection. Cognitive intervention aimed at these emerging functions might be more effective in the long run than a simple drill of math operations that lack the underlying cognitive basis.

Implementation of Vygotskian sociocultural theory in the classroom is based on the concept of learning activity. Sociocultural theory makes an important distinction between generic learning and specially designed learning activity (LA). Formal learning becomes a dominant form of child's activity only at the primary school age and only in those societies that promote it. Generic learning, however, appears at all the developmental ages in the context of play, practical activity, apprenticeship, interpersonal interactions, and so on. In a somewhat tautological way, specially designed LA can be defined as a form of education that turns a child into a self-sufficient and self-regulated learner. In the LA classroom, learning ceases to be a mere acquisition of information and rules and becomes learning how to learn. Graduates of the LA classroom are capable of approaching any material as a problem and are ready to actively seek means for solving this problem. Three elements constitute the core of LA: analysis of the task, planning of action, and reflection. Although analysis and planning feature prominently in many educational models, reflection as a central element of the primary school education may justifiably be considered a "trademark" of the LA approach. According to Russian Vygotskians (Zuckerman, 2004), there are three major aspects of reflection to be developed in the primary school: (1) ability to identify goals of one's own and other people's actions, as well as methods and means for achieving these goals; (2) understanding other people's point of view by looking at the objects, processes, and problems from the perspective other than one's own; and (3) ability to evaluate oneself and identify strong points and shortcomings of one's own performance. For each one of the aspects of reflection, special forms of learning activity were developed.

Chapter 4 shows how the development of rigorous mathematical thinking benefits from the use of the concepts of mediated learning and cognitive functions developed by Feuerstein et al. (1980). Feuerstein et al. postulated that mediated learning experience (MLE) reflects a quality of interaction among the learner, the material, and the human mediator. The quality of this interaction can be achieved only if a number of MLE criteria are met. Among the most important of these criteria are intentionality and reciprocity of interaction, its transcendent character (i.e., having significance beyond a here-and-now situation), and the mediation of meaning. Studies that follow this paradigm focused predominantly on the impact of MLE on the child's formation of cognitive prerequisites of learning and on the consequences of the absence or insufficient amount of MLE for the child's cognitive development.

The RMT theory purports that cognitive processes are formed through the appropriation, internalization, and utilization of psychological tools through the application of the MLE interactional dynamic. It is here that the RMT theory is informed by the unique synthesis of constructs from Feuerstein's theory of MLE and Vygotsky's sociocultural theory, particularly with regard to his emphasis that cultural symbolic artifacts become mediators of higher order cognitive processes. Vygotsky insisted that this process takes place through transformation of natural psychological functions into higher level culturally oriented psychological functions. For this process to be effective, the appropriation and internalization of these symbolic devices should be accomplished through the application of the three central or universal criteria of MLE – (1) intentionality/reciprocity, (2) transcendence, and (3) mediation of meaning.

One of the primary roles of MLE is to guide and nurture students to construct and internalize cognitive functions forming prerequisites of efficient learning activity. In the RMT paradigm these cognitive functions provide the foundation for and generate the mechanisms of rigorous thinking that become catalysts and building blocks for concept formation. We believe that students develop these cognitive functions through the appropriation, internalization, and use of psychological tools. A cognitive function is a specific and deliberate thinking action that the student executes with awareness and intention. There are two broad aspects of a cognitive function – the conceptual component and the action component – that work in relationship to each other to provide the cognitive function with its integrity as a distinct mental activity or psychological process. Embedded in this description is the notion that every cognitive function has a structure/purpose or structure/function relationship.

The conceptual component provides a "steering" mechanism to the mental activity by defining or giving description to the nature of the action that is taking place when the function is executed. For example, the cognitive function of comparing conceptually involves similarities and differences between two or more items. The action component of comparing is the mental action of looking for or searching for the attributes that the items share or have in common and those attributes that they do not have in common. In other words, comparing is the mental act of carrying out a search between or among two or more items that is guided by an identification of similar and different attributes the items possess.

These two broad components of a cognitive function give it specificity or distinction while lending it the capacity to intimately network, operationally, with other functions. For example, while comparing demands the forming of relationships and vice versa, the two cognitive functions are distinct and different. It is this contradistinction in nature that provides the foundation to the mechanism underlying concept formation through cognitive processing, supporting the notion that cognitive functions are tools of conceptual development.

The Feuerstein et al. (1980) *instrumental enrichment* (IE) cognitive intervention program offers one of the richest sources for the acquisition of symbolic tools and operations associated with them. The program demonstrated its effectiveness in significantly improving problem-solving skills in learning disabled, underachieving, and culturally different students (see Kozulin, 2000). The IE program includes 14 booklets of paper-and-pencil tasks that cover such areas as analytic perception, comparisons, categorization, orientation in space and time, and syllogisms. These booklets are called "instruments" because they help to "repair" a number of deficient cognitive functions.

Essential cognitive functions or specific thinking actions needed to construct any standards-based mathematical concept can be systemically developed through the IE program. This systemic development is promoted by three factors. First, the content of each of the 14 instruments is designed to support the construction of each of these cognitive functions. Although the instruments and their pages are different with regard to appearance of stimuli and/or levels of complexity or abstraction, each page practically provides the opportunity to deepen the construction of each cognitive function. For example, essential cognitive functions to start building and deepening conceptual understanding of variable and functional relationships between variables are conserving constancy, comparing, analyzing, forming relationships, and labeling. Each of these cognitive functions must be mediated to students to perform the tasks in instruments or units of tasks such as "Organization of Dots," "Orientation in Space," "Analytic Perception," and "Numerical Progressions."

A second factor is that the organization of the IE material and the activities are designed in such a way that any single task in one unit is related to the whole system of tasks in that unit. For example, all tasks of the "Organization of Dots" unit is of the same nature – an unorganized cluster of dots must be investigated to determine how to organize them by projecting virtual relationships. Each task in this unit requires analyzing a set of models that must be appropriated as psychological tools to compare and form relationships to carry out these projections. Each set of models is different on each page and progresses in complexity from the first page to the last page. When students practice use of the cognitive functions through these progressive levels of rigor the robustness of the cognitive functions is systemically developed.

A third factor that leads to the systemic development of cognitive functions through the IE program is that mediating students through the structure of a unit of tasks demands an organized approach that leads to the discovery of general cognitive principles and strategies. This element contributes to the development of theoretical thinking in students.

One of the better documented successes of the IE program is its ability to help culturally different students to acquire symbolic tools and learning strategies that were absent in their native culture but are essential in the modern technological society. From the foundational studies of Feuerstein et al. (1980) with immigrant students from North Africa to more recent research with immigrant students from Ethiopia (Kozulin, 2005a) it has been demonstrated that students' psychological functions are highly modifiable and can be radically transformed through the application of the IE program.

Chapter 5 demonstrates the "mechanics" of creating rigorous mathematical thinking through combination of Feuerstein's IE with fostering in students the development of mathematically specific psychological tools. Although mathematics is indeed the study of patterns and relationships, the need for very high levels of generalization and abstraction that are characteristic of this subject demands the employment of higher order psychological tools that are specific to the mathematics culture. The problem stemming from the traditional treatment of the subject, particularly with regard to formal instruction, is the failure to consider such devices or mathematically specific psychological tools as artifacts separate and distinct from mathematical content and with the instrumental functions of expressing, manipulating, transforming, and elaborating on such. Mathematics as we know it today is a cross-cultural synthesis that has evolved through a long, complex infusion of psychological tools and their cultural-historical significance originating from a number of cultures. Some of these tools are place value systems, number line, table, x-y coordinate plane, and equations.

A key theoretical construct in RMT is that each mathematically specific psychological tool has its unique design or structure and that this structure dictates the purpose, use, or function of the tool. In a general sense, each mathematically specific psychological tool has two roles. It activates, organizes, and integrates the use of specific cognitive functions to build a progressive understanding of both the mathematical operations and subconceptual elements needed to construct the mathematics concept, and it integrates subconceptual elements into a systemic matrix for the unified conceptual understanding.

For example, the structure of a number line stems from linear space that has been analyzed into equal-sized segments. Each segment is assigned the same quantitative value. The alignment of these sequenced segments organizes quantitative values into sequenced part/whole relationships that are sequentially labeled with numbers or other symbols that express this system of relationships. In terms of its role with regard to conceptual understanding in basic mathematics and in algebra, a number line can be used to organize the values of a variable into quantitative relationships and sequence, compare, and form relationships among whole numbers (counting numbers), natural numbers, fractions, rational numbers, and real numbers. A supportive role of the number line as a tool for conceptual understanding is the activation and organization of cognitive functions into specific clusters to promote the manipulation and integration of subconceptual elements of each concept.

Cognitive conceptual construction in RMT demands the concomitant use of cognitive functions, mathematically specific psychological tools, and previously developed subconceptual elements. Performing a structural analysis and an operational analysis of each standards-based concept identifies the most basic subconceptual element and progressively organizes and scaffolds other subconceptual elements to the most advanced element of the big idea. These analyses provide the blueprint for designing mathematical learning activities that will later guide students through a series of scaffolding processes, through the MLE, that help them to build their understanding of the concept.

The following activities with IE pages are mediated to the students: (1) define the problem (figure out what had to be done) on each IE page; (2) carefully analyze each psychological tool on the page to precisely define its critical attributes; (3) determine the relationship between the use of the tool and solve the problem to initiate the process of appropriating the tool according to its structure/function relationship; (4) utilize the tool to perform the IE tasks on the page; (5) identify and define the cognitive functions being used and how they are being used specifically to perform the tasks; (6) share and reflect on different strategies that were used, challenges encountered, and ways in which these challenges were addressed; and (7) apply psychological tools and the emerging cognitive functions to discover underlying principles connected to some tasks.

Mediating transcendence involves guiding students through worksheets that were specially designed by the authors to engage students in conscientiously practicing formation of conceptual elements of a mathematical concept by the joint use of psychological tools and cognitive functions. For example, for the concept of mathematical function, these conceptual elements are (1) change within the context of conserving constancy, (2) changeability, (3) interdependence, (4) cause/effect relationship, (5) variables, (6) functional relationship between variables, (7) independent and dependent variables, (8) one-to-one correspondence, (9) ordered pairs, (10) slope, (11) x and y intercepts, and (12) mathematical concept of function.

Much of the stimuli of IE tasks themselves embody patterns and relationships that readily lend themselves to mathematical expression, particularly with regard to big ideas in mathematics. In addition, it was determined that selected IE tasks contained foundational structures of mathematically specific psychological tools. This discovery and insight, when interfaced with the systemic nature of cognitive function development given in the previous section, has led to our formulation of systemic mathematical concept formation through rigorous mathematical thinking. For example, in the case of the concept of mathematical function, each page of the unit "Organization of Dots" can be used to structure activity for students to define the variables "proximity of the dots" and "overlapping of figures" and describe the functional relationship between these variables. In the unit "Orientation in Space" students can engage in identifying the variables "orientation of the boy" and "relationship between the boy and the object" and describing the relationship between these variables and so on. The units "Numerical Progressions" and others present stimuli and activities to build the understanding of slope as the change in the amount of the dependent variable (the effect) as a result of a specific change in the amount of the independent variable (the cause), and so on. As cognitive functions needed to construct a mathematics concept are systemically constructed through IE development, mathematically specific psychological tools such as a number line, table, and *x-y* coordinate plane can be appropriated directly from selected IE pages and the cognitive functions and tools together can be utilized to manipulate, organize, and form relationships among the patterns in the IE stimuli to systemically construct mathematical concepts.

Chapter 6 presents the RMT classroom application format and provides examples of successful application of the RMT model with various populations of learners. The application format for mathematics concept formation through RMT involves three factors – concept or topic, grade level, and time of application. Examples of the amount of time required for RMT classroom teaching to develop and/or improve understanding and skills with regard to specific mathematical concepts and topics are provided.

Empirical evidence of RMT training and teaching in classrooms involves students' change in disposition, cognitive development, and standards-based conceptual understanding and teachers' change in beliefs, instructional delivery, and views of mathematics curricular and student learning. First we present evidence of student change in classes of various cultures and/or groups in terms of pre- and postintervention results, analyses of students' mathematical activities, writings and reflections in RMT journals, student interviews, and some videotaped classroom sessions. After that we present results from chronicled observations of teacher practice made by RMT coaches, analyses of teacher questionnaires and interviews, examinations of teachers' lesson plans and their planning process, and examinations of how teachers analyzed student work and valued student oral and written responses.

One of the studies was conducted in the 4th-grade classrooms in a medium-sized Midwestern city in the United States. A class consisting of low-performing white, African American, and Latino students was taught the math concepts of fractions and function for 60 hours over a period of six weeks by a teacher in RMT training being coached by a RMT expert. During the same period two other 4th-grade classes of similar sociocultural and academic status within the same school were taught the same concepts by regular teachers. Postintervention cognitive ability test results for the RMT class were significantly higher than pretest results, whereas the gain scores in cognitive ability and on a standards-based test on function were statistically higher than gain scores for the two non-RMT classes. The RMT class scored statistically higher on the state's standards-based six-week math benchmark test, which was on fractions, than the two non-RMT classes. Student artifacts and journal writings are presented that demonstrate student learning.

In another study a 7th-grade class of African American students was taught the math concept of function for 16 hours over a two-month period by a teacher in RMT training being coached by a RMT expert. Statistically significant change in cognitive ability was produced while the gain score on the standards-based test on math function was statistically higher for this RMT class as compared to that of comparison group in a different school in the same community. Artifacts are presented for one student in the RMT class who generated mathematical insight while engaging in the full cycle of mathematical inquiry – representation, manipulation, and validation.

In yet another study the RMT model was applied with high school dropouts who had previously failed and/or "hated" mathematics. The RMT intervention was part of a larger six-month training aimed at equipping unemployed, "high-risk" inner-city residents with construction and environmental remediation skills. The goal was to increase employment while revitalizing dilapidated housing and reducing toxic environmental contaminants from inner-city communities. A statistical gain in cognitive ability was measured from pre-/posttesting while extensive evidence of conceptual change in mathematics and science was documented through chronicled student work and videotaped sessions. Students engaged in the full cycle of mathematical investigation – representation, manipulation, and validation – produced their own mathematical models and deepened their understanding of velocity, acceleration, gravity, force (including centripetal and centrifugal), the notion of relativity, and so on. In separate applications students derived structure/function relationships and functional relationships among variables during field investigations at a local planetarium, science and industry museum, natural history museum, butterfly haven, aquarium, and linear accelerator.

There is also evidence that not only students but also teachers participating in the RMT experience change their perception of mathematics culture. The change included transition from content-bound, algorithmic instructional delivery approaches to mediated process-driven conceptual teaching that proactively attempts to engage all students in thoughtful learning.

We conclude by stating that theoretical analysis as well as practical application of the RMT model in the U.S. multicultural classrooms confirms our main thesis that the constructive integration of the concept of psychological tools with the principles of mediated learning is capable of generating significant change in the students' mathematical reasoning. The RMT model may thus become an embodiment of the practical way for achieving the goals of standards-based education.

1 Culture of Mathematics

Sociocultural Needs Systems

The culture of mathematics has emerged from sociocultural needs systems over centuries. A needs system is a set of internalized habits, orientations, and predispositions that work together to provide the "blueprints" for human actions and the meanings for developing know-how. Sociocultural needs systems are integrally and functionally bound to the life and "ways of living" of the human society. They include both affective and cognitive-operational dimensions. These needs systems are engendered and shaped by a complexity of environmental, social, and cultural factors. The formation of such systems evolves into structures of meaning that carry the imprint of the society.

Different societies and different individuals may have different habits, orientations, and predispositions, and yet we are not aware of any human group that does not have some of the following needs systems: spatial and temporal orientation; identification of structure and function; part and whole relationships; change, constancy, and steady states; order, organization, and systems; balance, continuity, and symmetry; abstraction; and need for rigor. These needs systems reflect the objective circumstances of human existence, they are represented in human cognition, and they constitute the basis of mathematical knowledge.

One can approach these systems from three different perspectives: (1) the cultural-historical perspective, reflecting the developing needs of a given society; (2) the individual perspective, reflecting human genetic and biological predispositions on the one hand and individual appropriation of sociocultural tools on the other; and (3) the perspective of mathematical culture whereby these need systems were transformed into specific systems of mathematical meanings and operations. Usually each of these perspectives is explored in a different academic discipline, often completely disassociated with the others.

For example, cultural anthropologists may explore how the order and organization appear in the kinship system of a certain "traditional" society (e.g., Indian tribes of Amazon region; see Levi-Strauss, 1969). Psychologists will treat the child's ability to use the concept of order and organization as a cognitive function based on child's genetic endowment and maturational process (Piaget, 1947/1969), whereas mathematics educators, in their turn, will place the notion of order and organization into the context of specific mathematical curricular material (National Council of Teachers of Mathematics, 2000).

Our contention is that all three perspectives should be taken into account for better comprehension of the learning processes taking place in mathematics classrooms, especially under current conditions of multicultural education. Children come to these classrooms not as a "tabula rasa" but with a rich collection of notions and experiences informed in part by their cultural background and everyday life events and in part by their cognitive functions, both already formed and emergent. The task therefore is to be able to transform these preexistent elements of children's thought and understanding into the cognitive processes corresponding to contemporary mathematical culture. This is because although mathematics as we know it today represents an integration of elements from a number of cultures, it has its own unique culture that is distinctively different from "everyday ways" of doing things in various societies and cultures. Moreover, the cognitive functions observed in children in one sociocultural group appear to be naturally following the maturational path, but once we start observing children belonging to a different group we immediately see the culturally constructed nature of these skills (Rogoff, 2003). Thus one cannot take for granted a certain type of cognitive development in students of a multicultural classroom. Their cognitive functions, of both a general and more specifically mathematical nature, should be actively constructed during the educational process.

Historical Evolvement of the Sociocultural Needs Systems

During the Stone Age, ca. 5,000,000 to 3,000 B.C., the first societies were people who migrated in small groups as hunters of small game and gatherers of fruit, nuts, and roots (Eves, 1990). There is little doubt that the need for the groups to survive initiated and cultivated human activity that brought about orientations and dispositions that started the evolution of the sociocultural needs systems that would be essential to mathematical thinking. It is conceivable that the constant movement of groups in search of food, whose availability was related to seasonal and climatic variations, fostered individual and societal needs to deal with change and constancy and to develop spatial and temporal relations. Although there was not much time for self-reflection, daily existence fostered some sense of self-identity within the contexts of personal, group, and environmental transitions. Having a sense of self, perceiving others, and experiencing the regularity of the rising and setting of the sun, for example, created a need for and the conceptualization of constancy. However, growth in the individual and the members of the group, variations in the size and positions of one's shadow during the course of a day, and shifts in scenery and weather, for example, produced the need for and conceptualization of change.

The need for spatial dimensions and relationships grew out of a more immediate need to organize family and tribe members, objects, and events. Emergent requirements were the establishment of both a relative internal stable system of reference and an external absolute system of reference. These needs systems were bound to the mental operations of conservation of constancy in the context of dynamic change, seriation, integration, differentiation, interiorization, representation, and relational thinking.

The needs system for temporal orientation was probably initiated by the ongoing requirement to organize and sequence people, objects, and events while dealing with the past, present, and the future. Internalization of this needs system was probably enhanced by the underlying processes of tradition and culture that evolved as systemic vehicles to connect individuals and the group with the past, create meaning during the present, and plan for and anticipate the future. The relevant mental operations were seriation, representation, and conservation of constancy in the face of dynamic change, relational thinking, and operational analysis.

The development of material tools from stone, wood, bone, and shell for hunting and food preparation was accompanied by a need for structure/function and part/whole relationships. Most social and cultural activity proceeded through structure/function dynamics. Culture affected the internal and external worlds of the human "self," creating structures required for processes of transmission and perpetuation. This blueprint of culture replicated a system of learned behaviors and consciousness in the individual "selves" belonging to a given culture. At the individual level a fundamental and critical element of this perpetuating blueprint was the shaping of the interaction of "self" with the "other" (see Mead, 1974). Self-identity and self-determination became emerging constructs from the interaction of "self" with the "other" and the interactions of the "others" with others, forming the larger sociocultural self of the group. The structural elements of culture were the emerging systems of meaning that served as vehicles for ensuring, maintaining, and expanding the productivity and vitality of society. If one connects this anthropological perspective with the operations essential to rigorous mathematical thinking one cannot but admit a close affinity between the needs systems of cultural-civilizational development and the emergent elements of mathematical reasoning.

The above description should not be taken as an indication that we consider the development of societal needs systems, human cognitive functions, and specialized mathematical knowledge as either identical or parallel processes. It would be not only simplistic but also outright wrong to think that the developments of geometry can be directly derived from the particular needs to measure plots of land or that stages of children's development of the concept of space can be neatly fitted into the progression of geometry as a field of knowledge. Their relationships are much more complicated. What we would like to assert, however, is that behind each of the simple mathematical operations performed in the classroom is "hidden" some actual needs system (for example, the need to establish part/whole relationships), that the ability to establish these relationships depends on students' general cognitive functions, and that mathematics as a system of cultural knowledge provides tools and language that when properly appropriated can shape the students' understanding of these relationships beyond those available in their everyday experience.

From Symbolic Thought to Mathematical Expression: Historical Footprints of the Mathematics Culture

Mathematical reasoning is a form of symbolic thought. This is how the importance of symbol was described by one of the pioneers of the psychological study of symbolic thought, a German-American psychologist, Heinz Werner (see Werner and Kaplan, 1984, p. 12):

> Now it is our contention that in order to build up a truly human universe, that is a world that is known rather than merely reacted to, man requires a new tool – an instrumentality that is suited for, and enables the realization of, those operations constituting the activity of knowing. This instrumentality is the symbol. Symbols can be formed for, and employed in the cognitive construction of the human world because they are not merely things on the same level as other existents; they are rather, entities which subserve a novel and unique function, the function of representation.

Rather than just representing things, symbols always contain the element of generalization. The word *tree* represents a general concept of "tree" and will designate a specific tree only if additional perceptual or verbal information is provided. Even the simplest forms of mathematical symbolization represent general forms (e.g., triangle), quantities, and relationships rather than singular objects.

It took a long time for mathematical culture to develop its system of written symbols. The earliest forms of symbolization were closely related to such readily available tools as, for example, fingers. Different fingers, their positions on both hands, and the joints assumed a symbolic function. For example, tens or multiples of tens could be represented by touching the thumb to certain joints of the fingers (Menninger, 1969). This apparently simple representation changed the nature of both objects counted and the tools of counting (fingers). The objects, let us say a flock of sheep, became abstracted along only one of its possible dimensions – their number; then this number became represented by an entirely different medium – combination of bent fingers and touched joints. Fingers in this process also became abstracted, because only certain of their features acquired a new meaning, whereas others, such as size, strength, and color of skin, became irrelevant. Moreover, this new meaning of fingers was representational: they started representing something that had no natural connection to them as a part of the body. Once people moved to finger-counting systems more sophisticated than a mere one-to-one correspondence between 10 fingers and 10 objects, they apparently also changed their mental representation of objects and operations with them. They started thinking about countable quantities in terms of the available symbolic system of finger counting. Finger counting became one of the earliest human psychological tools that affected human thinking, memory, and representational abilities (Vygotsky and Luria, 1993).

Our next signpost on the road to mathematical culture is the abacus. The term designates two essentially similar but physically different counting tools. The Chinese abacus – probably the oldest of counting tools still in use – is a rectangular frame with beads on wires. The number of wires, their orientation (vertical or horizontal), and the number of beads vary from one type of abacus to another. The "European" abacus, probably of Shumerian origin, is a counting board with parallel lines on which pebbles or counters can be shifted. The Chinese abacus and its different modifications, such as Japanese *shuzan* and *soroban* or Russian *shchety* had been in continuous use for more than 2000 years. In Europe counting boards were used by ancient Greeks and Romans, became neglected during the early Middle Ages, and were then "rediscovered" in the 10th century. They soon became a standard feature of European educational and commercial institutions. It is claimed that the English Exchequer stopped using the counting table for tallying tax payments in only 1826 (!) (Pullan, 1968).

The idea behind calculation using an abacus is that one can "record" numbers using the positions of beads or counters on different lines or wires. The wires represent value, whereas the number of beads on a given line represents quantity. So to "record" the number 124, one moves one bead on the "hundreds" wire, two beads on the "tens" wire, and four beads on the "units" wire. By moving beads on relevant wires one can then add or subtract even very large numbers without taxing one's memory. Some types of abacuses have a divider that separates the column of beads representing units, tens, and so on from the column of beads representing the multiples of 5. In this way the operation of "borrowing" becomes even easier.

For us the abacus is important because it provides a very clear example of the connection between the function of symbolic representation and the beginning of mathematical culture. The abacus provides a two-dimensional symbolic space where physically similar beads acquire different quantitative meaning depending on their position on a given wire and the position of this wire on the frame. The arithmetic tasks that otherwise would require enormous concentration of attention and direct memory can be thus performed by using this symbolic apparatus and "recording" the intermediate results through positions of beads. The abacus not only helps to solve the practical task of calculation but also changes the type of cognitive operations involved. Whereas mental arithmetic requires direct memorization and operation with quantities, counting with an abacus shifts the focus from direct memory and processing capacity to understanding the rules and the use of algorithms. This is what Vygotsky and Luria (1993) had in mind when they considered symbolic tools as responsible for transformation of natural cognitive functions into higher level cognitive functions that are based on symbolic mediation. Cognitive transformation does not end with the acquisition of the abacus as an external symbolic tool. Skillful users of the abacus internalize it by creating its inner mental image and then start using this inner mental abacus as a calculation tool. In some societies, such as in Japan, the use of a mental abacus has become a cultural phenomenon that has gone beyond the practical needs of calculation (Hatano, 1997). There are afternoon abacus schools and abacus competitions, and there is even a system of ranks in the mastery of abacus skills.

There is no agreement among researchers whether the skillful use of the abacus as compared to paper-and-pencil calculations has a greater impact on the broader mathematical or general cognitive skills of students (see Stigler, Chalip, and Miller, 1986). In the context of our inquiry, however, such a comparison is not that important because both abacus use and written calculations are just two forms of symbolic representation and operation that have shaped the mathematics culture on the one hand and changed human cognition on the other.

The next symbolic tool to be discussed is a coordinate system. The history of this tool allows us to identify the moment when certain symbolic representation leaves the sphere of everyday activity and becomes developed into a mathematically specific tool. In this we see a rather marked difference between the abacus and the coordinate system. Both of them started as a practical means of calculation or spatial analysis, respectively, but although the abacus remained a tool of everyday counting without becoming a part of mathematical theory, the coordinate system made this transition.

Ancient Egyptian surveyors, in analyzing and organizing towns and open land into blocks and tracts, demonstrated a need of coordinates (Smith, 1958, pp. 316–317). Eves (1990, pp. 38, 47, and 346) reports that the Egyptians and Babylonians (3000–525 B.C.) developed the underlying mathematics to create a basic surveying and engineering practice to design and construct irrigation systems and parcel land, for example. Smith (1958, p. 316) points out that "the districts (*hesp*) into which Egypt was divided were designated in hieroglyphics by a symbol for a grid." Although there appears to be no literary account of the description of the coordinate system used by the ancient Egyptians, this hieroglyphic symbol has essential structural elements of the current Cartesian coordinate system.

The Romans also used the idea of coordinates in surveying and the Greeks used it in mapmaking. Hipparchus (140 B.C.) utilized longitude and latitude to locate positions on the surface of the Earth and in the heavens and located the stars by use of coordinates. From the ancient Egyptians to the Greeks and the Romans, surveyors and geographers located points by use of coordinates.

Apollonius (ca. 262–190 B.C.) gave evidence of the first phase of the mathematical development of coordinates (see Boyer, 2004, p. 46; Eves, 1990, p. 346; Smith, 1958, p. 318). By constructing auxiliary lines as axes for a previously given curve, he demonstrated first the perception of a function for a needed method and then the development of a structure to carryout that function (see Boyer, 2004, p. 46). Oresme (ca. 1323–1382) advanced the mathematical development of the coordinate system by reversing the process when he graphed a series of points that had uniformly changing *longitudines* (modern abscissas), the independent variable, and *latitudines* (modern ordinates), the dependent variable (see Boyer, 2004, p. 46; Eves, 1990, p. 346; Smith, 1958, p. 319). Oresme seems to be the first to demonstrate the instrumental nature of a system of coordinates as a mathematically specific tool when he first established a coordinate system and then plotted the geometric curve, taking a course of action from structure to function (see Boyer, 2004, p. 46). Smith

(1958, p. 319) points out, however, that there was a lack of continuity in Oresme's point systems.

At the same time it appears that the use of coordinates as originating in the work of ancient Egyptians, Babylonians, and Greeks had not been generalized as a methodological tool to support mathematical processing. Whereas the ancient Greeks studied curves and algebraic relationships with coordinates, their approach was quite different from that of Fermat (1601–1665) (Boyer, 2004, p. 75). The former started with the subject of study and moved to formulating the algebraic relationships by constructing the semblance of a coordinate system as an intermediate step. When studying a given curve, they superimposed particular lines on the curve and from this association used rhetorical algebra to construct a verbal description of the properties of the curve. The set of particular lines served as a coordinate system. Fermat reversed the process. His starting point was an algebraic equation followed by the construction of a coordinate system and then using such to define a curve as a locus of points (see Boyer, 2004, pp. 76–77). Coolidge (1942, p. 128) states, "Descartes showed that if any curve were mechanically constructible, we could translate the mechanical process into algebraic language, and find the equation of the curve."

Of course, it is clear that neither Fermat nor Descartes was the originator of coordinates; neither were they the first to utilize graphical representation. But although Fermat and Descartes did not invent coordinates, their combined work established the mathematical generalization that a given algebraic equation in two unknown variables determines a particular geometric curve, which formalized the need for a stable, generalized system of coordinates in the mathematics community. Leibniz (1646–1716) was the first to establish such in his letters in 1694 in which he gave equal weight to the two coordinates and used the terms *coordinates, abscissa,* and *ordinate* in the sense they are used today (see Boyer, 2004, p. 136; Eves, 1990, p. 352). Descartes' major discovery was the principle that the equation contained every property of the curve, provided there is a method to reveal it or provide logical evidence of its presence. In the same sense Leibniz's most significant achievement was the revelation that the differential coefficients were also dependent on the curve. Standing at the interface between proving a geometric theorem and an algebraic functional relationship between two variables, whether from a perspective of finite or infinitesimal dimensions, is the need for a structure to carryout a function. The Cartesian coordinate system as a mathematically specific tool meets this need.

The history of the coordinate system from its early origins in the work of the Egyptians, Babylonians, and Greeks and into its classical formulation by

Fermat, Descartes, and Leibniz teaches us how a graphic tool that started as an auxiliary means for surveying and mapmaking became a representational system linking two major areas of mathematical culture – geometric and algebraic. In the epoch of Fermat and Descartes each one of these areas has already developed into a field of theoretical knowledge with its own language, rules, and operations. Algebra and geometry no longer just established the connection between objects and their symbolic representations but created a network of internal relationships between mathematical expressions in each of these areas. In the work of Fermat, Descartes, and Leibniz the next step has been made: the relations were established between the languages of two relatively independent systems, thus creating an integrated mathematical theory. It is this synthesis that is often lacking in teaching mathematics is our schools. Students often perceive algebraic tasks and graphs as two absolutely unrelated entities each one composed of standard operations that, if properly chosen, will lead to a correct answer. It is little wonder that a simple task of finding the speed of a car when a graph of the functional relationships between the distance and time is given was correctly solved by only 54.3% of the 8th-grade students in the Third International Mathematics and Science Study (Smith, Martin, Mullis, and Kelly, 2000).

Mathematical Learning Activity

Mathematical activity seeks to make meaning from aspects of patterns and relationships through abstraction. The nature of mathematics demands a detachment and freedom from specific stimuli and objects in order to maintain compatibility with the growing body of knowledge it generates. This mathematical knowledge is generated and qualified by logic and creativity through a cycle of investigation that comprises representation, manipulation, and validation (American Association for the Advancement of Science, 1990, 1993). Thus, mathematical activity generates and qualifies mathematical knowledge through a process of inquiry that demands logical scrutiny and precision. It presupposes on the one hand the compliance with the principles of logical inference and on the other being useful by enhancing the body of mathematical knowledge and/or being uniquely compelling or intriguing.

The goal of mathematical learning is the appropriation of methods, tools, and conceptual principles of mathematical knowledge based on efficient cognitive processing that constitutes an essential prerequisite of mathematical learning. Not every activity that takes place in a mathematics classroom qualifies as a genuine mathematical learning activity. In this respect our paradigm of Rigorous Mathematical Thinking (RMT) is based on the extension of

Vygotsky's notion of learning activity and "scientific concepts" (discussed in Chapter 3) to the domain of mathematical classroom learning. Vygotsky and his followers made a rather sharp distinction between learning in a generic sense and a specially designed "learning activity." Learning in a generic sense takes place all the time and in all possible contexts. We learn when we play, when we work, when we are involved in interpersonal relationships, and so on. But in all the above contexts learning is not a goal but a means or a by-product. Only in a specially designed learning activity does learning become its own ultimate goal and objective. Through this activity we learn how to learn and gradually become self-regulated learners.

The above definition of learning activities assumes that, first, students should be provided with means of learning. This apparently trivial requirement is often neglected in mathematics classroom. Instead of helping students to develop mathematical learning skills, they are provided with mathematical facts and operations. It is often presumed that any normally developing child has "natural" learning skills to appropriate these facts and operations. We, on the contrary, claim that the relevant learning skills should be developed though specially designed activities. This leads us to the second point that effective learning presupposes activities. This again may sound trivial because, after all, considerable time in the mathematics classroom is dedicated to the activity of problem solving. However, presenting students with mathematical tasks and then checking the correctness of the results is not a learning activity. The learning activity includes *orientation* in the presented material, *transformation* of the presented material into a problem, *planning* the problem-solving process, *reflection* on chosen strategy and problem-solving means, as well as *self-evaluation.* All of the above elements are universal for any learning activity, and yet in each curricular area they are supposed to be attuned to the conceptual understanding characteristic of a given field of knowledge. In this sense the mathematical learning activity is expected to lead students toward the formation of "scientific" concepts as defined by Vygotsky and his followers (see Karpov, 2003a). The scientific concepts are *theoretical* in the sense that they capture the central conceptual principle of a certain phenomenon rather than provide a collection of empirical facts or skills. These concepts are *generative* because they allow us to predict or design all possible empirical manifestations based in the identified principle. And scientific concepts are *systemic* because their meaning is always derived from the system of relationships between different concepts.

According to Vygotsky's model, learning activity is expected not only to develop efficient learning in a given curricular area but also to promote the students' more general cognitive development. So here the relationships between the students' general cognitive functions and their curricular learning enter into genuine reciprocal relationships. On the one hand, curricular learning is based on the development of general cognitive functions, whereas, on the other hand, it should be designed in a way that further develops these functions. That is why the Vygotskian model of education is called "developmental education" (Davydov, 1988a, b, c). For example, the 1st-grade mathematics textbook developed by Vygotsky's followers in Russia and adopted in the United States (see Davydov, Gorbov, Mikulina, and Saveleva, 1999; Schmittau, 2003) is devoted primarily to the development of general learning and problem-solving skills and only at a much later stage introduces counting and numbers. Students who study mathematics according to this curriculum demonstrate later in their school life a much better facility with tackling "nonstandard" problems, transferring the principles to the new areas of applications, and so on.

It is thus important to elaborate on specific criteria for determining which actions in the mathematics classroom meet the RMT standard. (1) The classroom activity should aim at creating a structural change in the students' understanding of mathematical knowledge. A mere accumulation of information or skills does not qualify as a structural change. By structural change we mean a qualitative change whereby the student changes his or her level of mathematical comprehension. One may see here a certain parallel to Piaget's (1970) notion of cognitive structures that are characterized by systemic organization, self-regulation, and transformation. A new level of comprehension should have a systemic unity so that all elements become involved in it. In addition, the change should, on the one hand, be strong enough to withstand the temptation to solve novel tasks by reverting to previous, less advanced level of reasoning and, on the other hand, open enough for further transformations. (2) The classroom activity must aim toward, and therefore be a part of, a process for constructing a "scientific" mathematics concept, characterized by its theoretical, generative, and systematic nature. Increased efficiency with certain mathematical operations that remain isolated, atheoretical, and devoid of generative aspect cannot be qualified as responding to RMT goals. (3) The learning activity must introduce the students to the language and rules of mathematics culture. Whatever the students' native language, family customs, or everyday experiences, the joint activity in the mathematics classroom should provide them with an opportunity to be introduced to a culture new to all of them and common to all of them – the culture of mathematical thought. This culture has its own language, rules and customs, history, and challenges. In this sense every mathematical classroom is multicultural. However, unlike other multicultural classrooms, here there are no children belonging to the "majority" culture, because no child can claim mathematics as his or her native culture.

Mathematical Knowledge

Mathematical knowledge exists at three levels: mathematical procedures and operations, mathematical concepts, and mathematical insights. This knowledge may not originate in the everyday experience of the child, which, however, does not disqualify these experiences from being one of the sources for comprehending them mathematically. In this respect the RMT paradigm differs from the popular constructivist approaches that suggest starting with children's everyday experiences and then attempting to create a conceptual change (Fosnot, 1996). We, on the contrary, suggest starting with the construction of a new mathematical *subject* and then applying the methods, language, and operations characteristic of this subject to the everyday experiences of the child. Mathematical operations involve basic processes of organizing and manipulating mathematical information in meaningful ways that support and build important ideas and concepts. As mentioned earlier all mathematical concepts are "scientific" according to Vygotsky's definition of this term, that is, they are theoretical, systemic, and generative.

Knowledge in general consists of a body of facts, concepts, principles, and axioms that are understood or grasped by the mind. In the RMT paradigm we emphasize the interconnections among mathematical knowledge, mathematical learning activity, and development of students' higher cognitive functions. Mathematical knowledge, therefore, should be perceived by the student as emerging from his or her mathematically specific learning activity supported by cognitive functions. The RMT paradigm removes the artificial separation between higher cognitive processes as belonging to the area of "cognition" and mathematical knowledge as belonging to the area of "curricular content." Mathematical knowledge taken in its conceptual form becomes a part of students' higher form of cognition – conceptual reasoning. As stated, mathematical knowledge consists of organized, abstract systems of logical and precise understandings about patterns and relationships. Patterns consist of repeating aspects of objects and events. When the conservation of constancy and change in these phenomena are investigated, described, defined, represented, manipulated, validated, made quantifiable, and generalizable the birth of mathematical knowledge takes place.

In the RMT paradigm, specific well-defined cognitive processes drive mathematical operations and procedures. Cognitive processes specific to the mathematical domain include quantifying space and spatial relationships, quantifying time and temporal relationships, projecting and restructuring relationships, forming proportional quantitative relationships, mathematical inductive-deductive thinking, elaborating mathematical activity through cognitive categories, and so on. Cognitive processes facilitate the organization and manipulation of mathematical knowledge to make meaning through the usage of mathematical operations and procedures. These procedures also play a critical role in deepening and expanding students' understanding and their ability to build mathematical concepts and insights. Mathematical insight is derived from one or more conceptual understandings, forming relationships between or among these understandings, and constructing new ideas and/or applications.

Mathematical knowledge is not just discrete ideas, operations, and procedures. In the RMT paradigm students are made constantly aware of the structural and systemic nature of this knowledge that forms an interdependent hierarchy, beginning with the most core concepts and expanding to the big mathematical idea. No subject is taught in isolation; no concept appears as separate or as self-contained. The systemic nature of mathematical knowledge is acknowledged by many didactic approaches. What distinguishes RMT is our emphasis on concomitant systemic organization of mathematical knowledge, learning activities, and emergent higher cognitive functions.

Need for Mathematically Specific Psychological Tools

Mathematically specific psychological tools extend Vygotsky's (1979) notion of general psychological tools. According to Vygotsky, the transition from the natural cognitive functions to the higher cognitive processes is achieved through the mediation of socioculturally constructed symbolic tools. Symbolic devices and schemes that have been developed through sociocultural needs to facilitate mathematical activity when internalized become students' inner mathematical psychological tools. Mathematically specific psychological tools play an instrumental role in both mathematical activity and the utilization of cognition in constructing and applying mathematical knowledge.

Basically there are four categories of mathematically specific tools that have the potential of becoming students' inner psychological tools: (1) signs and symbols, (2) graphic/symbolic organizers, (3) formulas and equations, and (4) mathematical language. The instrumentality of such tools stems from core conceptual elements of the mathematics culture, such as quantity, relationships, abstraction/generalization, representation, precision, and logic/proof and their unique structure/function relationship. The structure/function relationship of such tools appears differently depending on the category of the tool. Signs and symbols in mathematics present structure through the required encoding (putting mathematical meaning or significance into a code or symbol) and decoding (pulling mathematical meaning or significance from the sign or symbol). Thus, there is symmetry in this structure. This structure carries out the function of capturing a mathematical conceptual element while activating and using cognitive functions such as labeling-visualizing, comparing, forming relationships, and conserving constancy. The integration of a conceptual element with cognitive functions creates a mathematical semantics that is connotative in nature. Combining the use of signs and symbols based on the rules and logic of the mathematics culture is formed through a mathematical syntax.

Although all mathematical signs and symbols carry the function of encoding-decoding, many have roles that extend beyond this level and form rather complex relationships. On the one hand, for example, the symbol for infinity (∞) is restricted to capturing the concept of endless or limitless quantity of some element such as space or time. On the other hand, for example, the symbol for summation ($\sum$) not only captures the concept of summation of quantities but also extends to the notion of forming relationships by composing or integrating quantities of the same dimension or unit as defined by the formulation that is designated on the right of the summation sign. Some or all of these quantities could be represented by variables, either of which may be determined through another defined relationship.

Although the use of symbols and signs to encode-decode is arbitrary, much of symbolic use in mathematics is determined by consensus of the mathematics community. Though specific signs and symbols have nothing to do with the critical attributes of the "object" or process they encode, there are some agreed-on customs of the use of specific symbols. For example, X usually designates a variable. Such permanency in the use of a sign or symbol to always encode-decode the object or action may become an obstacle for student learning in the mathematics classroom. Students often perceive a symbol or sign as bearing the essence of the object or event it is representing. When asked whether a variable can be encoded by a letter T, they would respond that T designates time, whereas a variable should be designated by X. Because students start learning mathematics by using the product of what mathematicians have developed through mathematical activity, the sequence of mathematics learning by students is the opposite of theoretical mathematics construction. Even the arbitrary use of symbols to represent, for example, a variable in algebra is perceived by students in the classroom to carry the essence or critical attributes of its representation.

In certain cases the role and meaning of the sign or symbol can be determined only from contextual usage. For example, the symbol of a dot (.) is used in the expression 3.4578 to separate the one's place value from the tenth's place value, whereas in 3.3333 . . . . the succession of dots (. . . .) encodes a limitless continuation of the digit 3. In $3 \cdot 7$ the "same" symbol encodes the operation of multiplication, whereas on a number line and *x-y* coordinates it represents the origin. Thus, the structure/function relationships of mathematical signs and symbols are governed by the culture of a mathematical semantic field.

This field encompasses the properties of quantities the symbols represent and their arithmetic operations, some of which are illustrated below:

1. Additive identity: $p + 0 = p$ and $0 + p = p$; adding zero to a quantity gives the same quantity.
2. Additive inverse: $p + n = 0$ because in general the quantity $-p$ is the unique solution for the quantity n.
3. Arithmetic negation: $m - h = m + (-h)$; subtracting a quantity is the same as adding its opposite.
4. Multiplication and negation: $-p = (-1) \times p$ and $p = (-1) \times (-p)$; negation is multiplying a quantity by -1.
5. Opposite of opposite: because the opposite of the opposite of a quantity is the quantity itself, in general, $-(-p) = p$.
6. Commutativity of addition: $p + m = m + p$; the order of two quantities does not affect their sum.
7. Associativity of addition: $(p + m) + c = p + (m + c)$; when given three or more quantities to be added, it does not matter whether the first pair or the last pair is added first.
8. Commutativity of multiplication: $p \times q = q \times p$; the order of the two factors does not affect the product.
9. Associativity of multiplication: when multiplying three or more quantities, it does not matter whether the first pair or the last pair is multiplied first.

The next category of mathematically specific tools is graphic/symbolic organizers. The first in this category is a number system with its place values. In the RMT paradigm a number represents quantity, amount, or value and is originally derived from measurement. It is this derivation that has produced insight into the need for a comprehensive, interrelated, coherent, and expandable system of artifacts with an organizing component to fully represent the idea of quantity. This derivation also brought about understanding that a single isolated quantity cannot exist and have meaning in and of itself because through measurement there is no way it can be produced as a single, independent item. The structure of this system is established from two components: the unit of quantity, called the base, and the analyzing-integrating aspects of the place values and their encoding-decoding role. The function of the number system is to organize, compare, and form relationships among quantities in a way that is logical and ensures the integrity of the process. Thus the structure of this tool leads to the important mathematical functions of understanding and manipulating quantities while integrating the use of cognitive functions with this process.

The number line is the second tool in this category. Its structure stems from analyzed linear space with series or levels of segments within segments. Each level of segments is equally segregated space to capture quantitative equivalency. The function of this tool is to identify, compare, analyze, integrate, and form relationships between and among quantities. The third tool in the category is a table. Its structure is derived from its columns and rows. The heading for each column organizes mathematical data or information into a set, whereas the left-to-right movement in one row from one column to the other forms a relationship between the two items. Continued movement to the next column forms a relationship between relationships. The entire set of relationships among relationships expressed in the table forms a functional relationship among the data. Thus it is clear that the function of the table develops mathematical conceptual understanding while organizing and orchestrating the use of cognitive functions.

The *x-y* coordinate plane is a tool that consists of two number lines that intersect at their origins and form right angles. Each number line is used to represent the quantities of a variable and while being used together they can function to bring about the formation of relationships between corresponding values of the two variables or ordered pairs. This can happen when the two variables are in a cause/effect relationship, an input/output relationship, or an independent/dependent relationship. When a set of relationships is established between these ordered pairs or corresponding relationships of values for the two variables, a functional relationship between the two variables can be visibly shown in the artifact. Internalization of this functional relationship will help students understand how these two variables are conceptually working together in their interdependency. Thus the *x-y* coordinate can assist in providing students with both a theoretical conceptual understanding of interdependent relationships in everyday life and technology and with the practical know-how of how to analyze many unknown situations, construct meaning from them, and formulate applications and innovations.

The next category of mathematically specific tools is formulae and equations. The structure/function relationship in formulae and equations demands both quantitative and conceptual equivalency. The structure for such is derived from the equal sign (=), encoding the notion that whatever exists on the left is equivalent to whatever exists on the right in total quantity. Each side is also expressing a mathematical or a group of mathematical concepts that must be equivalent with regard to the object and its unit or dimension. The two sides may employ different operations and procedures to bring about this quantitative conceptual equivalency. The expression of quantitative equivalency may be as simple as $A = B$ or $5 = k + 7$, where the object being conceived is more generalized. However, in formulae such as $f = m \cdot a$, $E = m \cdot c^2$, or $y = \frac{1}{2}(b \times h)$ and $A = (l \times h)$, the object's force, mass, acceleration, energy, speed of light, and geometrical area are conceptually specific and demand definite units or dimensions. In these cases the use of cognitive functions must be extensive and abstract.

The last category of mathematically specific tools is mathematical language. According to Vygotsky (1986), language occupies a unique place in human cognition, being at the same time a universal mediator between other cognitive functions and a higher cognitive process of its own right. Thus for the mathematical language there is a dual function – to express mathematical thought while at the same time serving as a medium for creating mathematical thought. Thus this tool serves the RMT student as both a vehicle and a superordinate cognitive function. The expression of mathematical thought is not only for those who listen but also for the learners themselves. Vygotsky (1986) was the first to emphasize the importance of students' inner speech for the development of their reasoning. The learner must engage in self-talk while listening, reading, composing, writing, reflecting, and so on. That is why in RMT classrooms students are encouraged to embody, both orally and in writing, the process of their comprehension and solution of mathematical problems. Mathematical language deals not only with the signs and symbols of mathematics but also specific mathematical concepts, the expression of the operations, the labels and meaning of the cognitive functions, and the structural/functional nature of all other mathematically specific tools. The dual function of mathematical language is to provide prerequisites for mathematical reasoning and to serve as a medium of students' mathematical reflection and self-expression.

Mathematical Rigor

Any genuine mathematical reasoning is rigorous. We define mental rigor as that quality of thought that reveals itself when students' critical engagement with material is driven by a strong, persistent, and inflexible desire to know and deeply understand. We now turn to the notion of rigorous engagement – a state in which three components of a learning interaction are mutually interacting with a synergy that appears to be self-perpetuating. The three components of the interaction are the teacher, the learner, and the task. The task in this context includes not only material to be learned but also the prior knowledge, experience, and culture of the learner and the teacher. The process commences, not at a level of ease or synergy, but with struggle brought about through some disequilibrium or dissonance between the learner and some aspect of the task. The development of rigor must be sustained by the teacher, who may have to struggle with the learner and the task to precisely identify and define the nature of a dissonance. The teacher's intention in an RMT classroom is to encourage and guide the learner into a psychological cooperative commitment. This commitment begins to nurture a level of trust between the learner and the teacher. This trust both positions the learner to not be afraid of failure and encourages teachers to reveal their own shortcomings, either naturally occurring or intentionally displayed. It is here where the teacher is inviting the learner to be a comediator – co-investigator, coteacher, and colearner – in structuring and maintaining the engagement. The learning material is thus transformed and takes on a life of its own.

When rigor is achieved, the learner becomes capable of not just engaging in specific problem solving but also of reflective thought. In a more general sense students learn how to function both in the immediate proximity as well as at some distance from the direct experience of the world. This all contributes to the development of students' metacognitive skills. We have described a cognitive function as having three components: (1) a conceptual component, (2) an action component, and (3) a motivational component. This metacognitive state means that the three components of the cognitive functions are taking on a role beyond the acquisition of mathematical knowledge but are serving to evaluate what is being accomplished and to plan, weigh options, predict, and select paths for further action.

Rigor describes the quality of being relentless in the face of challenge and complexity and having the motivation and self-discipline to persevere through a goal-oriented struggle. Rigor takes on the attributes of intrinsic motivation and task-intrinsic motivation. Rigorous thinking requires an intensive and aggressive mental engagement that dynamically seeks to create and sustain a higher quality of thought. Thus the learner is compelled to construct conceptual theoretical learning that produces principles beyond the content and context of the stimuli. Mathematical rigor is initiated and cultivated through mental processes that engender and perpetuate the need for rigorous engagement in thinking. Rigorous thinking can thus be characterized as sharpness in focus and perception; clarity and completeness in definition, conceptualization, and delineation of critical attributes; precision and accuracy; and depth in comprehension and understanding.

2 Goals and Objectives of Mathematics Education

About 40 years ago Jerome Bruner observed the following:

> I shall take it as self-evident that each generation must define afresh the nature, direction, and aims of education to assure such freedom and rationality as can be attained for a future generation. For there are changes both in circumstances and in knowledge that impose constraints on and give opportunities to the teacher in each succeeding generation. It is in this sense that education is in constant process of invention. (Bruner, 1968, p. 22)

The overarching goal of education in the United States is to prepare students to function as productive citizens in a highly industrialized and technical society. In such a society, technological developments and advancements are driven and promoted by mathematical and scientific discovery and application. Since the 1960s there have been numerous attempts to reform education so that it provides a greater focus on scientific and mathematical literacy. These attempts were designed not only to increase the number of professional scientists, mathematicians, and engineers but also to equip all members of society with awareness, skills, and understanding to perform more effectively in the context of such technological development and advancement. Therefore, this process is important at three levels: (1) to maintain personal membership and a sense of belonging to the general U.S. culture, which means possessing a sense of self-worth as a *bona fide* member of the society; (2) to have the basic skills and ability to perform everyday tasks, such as reading the newspaper, operating a home appliance, and creating and managing a budget; and (3) to increase the pool of highly qualified science, engineering, technology, and mathematics professionals who are U.S. citizens and are educated in U.S. schools.

The achievement of this overarching goal of education demands a system of high-quality schooling in mathematics and science from pre-K through the

12th grade. However, a major problem exists. The lack of rigor in the teaching of mathematics and science education in the United States is highlighted in a report entitled *Before It's Too Late* prepared by the National Commission on Mathematics and Science Teaching for the 21st Century (2000) and submitted to former U.S. Secretary of Education Richard W. Riley. This report presents a serious indictment of the quality of mathematics teaching from several perspectives. First, it provides evidence that mathematics teaching in classrooms across the United States is archaic and leads students through a dull routine that is neither exciting nor challenging. The report cited a study of videotaped sessions of 8th-grade mathematics classes (Olson, 1999) from the Third International Mathematics and Science Study (TIMSS). All lessons comprised the following: "(1) a review of previous material and homework, (2) a problem illustration by the teacher, (3) drill on low-level procedures that imitate those demonstrated by the teacher, (4) supervised seatwork by students, often in isolation, (5) checking of seatwork problems, and (6) assignment of homework. In not 1 of 81 videotaped U.S. classes did students construct a mathematical proof" (National Commission on Mathematics and Science Teaching for the 21st Century and submitted to former U.S. Secretary of Education Richard W. Riley, 2000, p. 12). A second perspective is that such poor-quality teaching will not produce the number of qualified workers demanded by the great increase in technologically, scientifically oriented firms that are rapidly developing. The report projects that such industries will by 2008 create approximately 20 million jobs in the U.S. economy. A third perspective is that poor-quality teaching in mathematics and science reduces the general population's ability to make responsible, informed decisions in everyday living and threatens the nation's security. A fourth perspective is that mediocrity in mathematics and science teaching undermines the needed mathematics and science knowledge that is essential to our culture and way of living.

United States leaders have been grappling with perceived problems in mathematics and science teaching since the 1950s with the Russian invention and launching of Sputnik 1 in 1957. The orbiting of Sputnik caused a national self-assessment of American education. The National Science Foundation, established in 1950, participated in the effort to examine the status of U.S. education, particularly in the area of science. Congress expanded the federal role in higher education with the passage of the National Defense Education Act of 1958. However, the focus on improving K-12 mathematics and science education remained limited until the 1960s.

Before his assassination, President John F. Kennedy attempted to expand federal influence on education. By 1965, the role of the U.S. government in precollege education dramatically expanded with the Elementary and

Secondary Education Act (ESEA) of 1965, spearheaded by President Lyndon B. Johnson. This legislation provided more funds for federal research and development and support for disadvantaged students. Concurrently, during the 1960s, the federal government was concerned about accountability for student academic achievement. Under the leadership of Francis Keppel, then U.S. commissioner of education (1962–1965), a bold attempt was made to influence Congress to approve policy for a federal assessment system. After Keppel's tenure and much controversy, the National Assessment of Educational Progress, under the trusted auspices of the Educational Commission of the States (ECS), was funded with a mixture of private and public funds. About the same time, in 1964, the International Association for the Evaluation of Educational Achievement (IEA) piloted a study. The First International Mathematics Study (FIMS) was conducted, which sought to identify critical factors in student mathematics achievement in various countries, including the United States. Following this study, a six-subject study was conducted that included science. In the 1980s the IEA conducted a second mathematics study (SIMS) in 20 countries and a second science study (SISS) in 24 countries.

In 1983, tremors of upheaval were felt throughout the educational community and the nation as a whole. Under the leadership of President Ronald Reagan, the National Commission on Excellence in Education published *A Nation at Risk.* This report shook the foundation of schooling practices and demanded high expectations in elementary and secondary education. The call was made for a transition from the tradition of behavioristic teaching practices to a more cognitively oriented education that acknowledges the social context of individuals and groups (classwide) and includes problem solving. The *Nation at Risk* report initiated the first wave of educational reforms that directly addressed the raising of standards in a variety of areas – academic content, assessment programs, and preservice teacher standards. However, the reforms of the 1980s were limited in their effectiveness. The fragmented and contradictory policies coupled with minimal change in curriculum, instruction, student learning, and achievement took reformers back to the table for further dialogue, argument, research, and development.

The second wave of education reform occurred in the 1990s and featured several changes in the ideology of how reforms in teaching and learning should be developed. Reformers of the 1990s created the idea of systemic change also labeled standards-based reform. This approach was composed of three parts: (1) challenging content standards that identified what students should know and be able to do, (2) aligning policy and accountability efforts to the content standards, and (3) restructuring governance systems that support standards-based education at the local district level.

This wave of reform efforts was informed by four documents. Beginning in the mid- to late 1980s, three sets of guidelines were published that provided substance for the dialogue around the need for national mathematics content standards. They were the California Curriculum Frameworks, the National Council of Teachers of Mathematics' *Curriculum and Evaluation Standards for School Mathematics,* and the National Research Council's *Everybody Counts: A Report to the Nation on the Future of Mathematics Education.* A fourth source, the American Association for the Advancement of Science (AAAS), also contributed to this reform dialogue in 1989 with the publication of its text on scientific literacy, including mathematical literacy. Close to the end of the 20th century various ideas about the purpose of mathematics and science education became a catalyst that gave birth to the standards movement.

A comprehensive vision of literacy in science, mathematics, and technology in the form of achievable learning goals in these disciplines was presented in *Science for All Americans* (American Association for the Advancement of Science, 1990) as the report of a 3-year collaboration of several hundreds of scientists, mathematicians, engineers, physicians, philosophers, historians, and educators. *Science for All Americans* presented a broad view of scientific literacy from three perspectives. First, it thematically focused on the interdependency of science, mathematics, and technology as human enterprises and collectively presented this interrelatedness as the science endeavor. A second perspective is consideration of science as both the natural sciences and the social sciences. A third and very important perspective is its emphasis on habits of mind that embrace not only knowledge acquisition but also the need for developing independent thinking and human values.

A companion report, *Benchmarks for Science Literacy* (American Association for the Advancement of Science, 1993), developed through a collaboration of 150 teachers and administrators, delineated the next step to reaching the goals recommended in *Science for All Americans* by specifying what students should know and be able to do in science, mathematics, and technology at various grade levels. These specifications serve as a standard curriculum to strategically inform educators, parents, administrators, and others across the country in planning inquiry-based education and instruction.

Current standards for mathematics and science education in the United States grew out of a series of reform efforts on local, state, and national levels that have extended from the 1980s to the present. However, the 21st century brought in a more radical approach to federal involvement in educational reform. The No Child Left Behind Act (NCLB) of 2001, signed by President George W. Bush, ushered in a new age of assessment and accountability measures for precollege education. The NCLB Act took bold steps to merge content, assessment, and accountability measures into one policy. This controversial legislation demanded that states take a hard look at their education reform policies and demanded alignment with NCLB directives.

The NCLB Act supplanted the ESEA and its funding mechanisms (Title I through Title IX). States were accountable to comply with the mandates of the Act to receive this funding. Local districts and its schools were required to be responsible for attaining specified yearly gains, maintaining a staff of "highly qualified" teachers, and utilizing "scientifically based" curriculum materials. All states were required to participate in NAEP testing for reading and mathematics at particular grade levels, in addition to the requirement of mandatory testing of 3rd through 8th graders in reading and mathematics.

It is questionable whether this latter approach would generate a genuine reform because its nature is bureaucratic rather than conceptual. School education in the United States seems to suffer from the same malaise that it had 50 years ago with the advent of Sputnik: inadequate student achievement as a result of inappropriate teaching and the lack of rigorous instruction and high expectations for student achievement. A reduction of state and local control to a dramatic increase of federal control and mandate does not seem to offer an adequate response to this situation.

The Standards Movement in Mathematics and Science in the United States

The term *standards-based concepts* refers to those academic mathematics and science concepts that are central to the national curriculum guidelines for mathematics and science education formulated through education reform efforts in the United States. Reys and Lappan (2007) conducted a national study of state mathematics standards. This study was designed to investigate the extent to which states had developed state mathematics standards that comply with national reform guidelines and how the quality of such development changed over time. First they pointed out that the mandates of the No Child Left Behind law brought about a shift in the reform effort. In a standards-based learning environment students are expected to become the center of the curriculum instruction/assessment process. Teaching begins with activating the learners' prior knowledge and experience. What is to be taught or learned is constructed from the individual student's and the class's rich experiential repertoire.

Reform efforts and the current standards in mathematics and science evolved out of a need to address poor student achievement in these subjects. Student achievement presumably is based on the acquisition and application of mathematical and scientific knowledge. This acquisition and application of knowledge is based on the quality of teaching and learning taking place in the classroom. Thus, an important question to be raised is: What are the relationships that exist among the standards, curriculum, mathematical and scientific knowledge, teaching, learning, and student achievement?

First, let us examine the relationship between the mathematics standards and a mathematics curriculum. The standards state what students should know and be able to do at a particular time in their matriculation in school. A mathematics curriculum provides a body of information, principles, concepts, rules, and so on, and some organized presentation of these to acquire mathematical knowledge. The curriculum is often a ready-made document that is passed to teachers as an instructional roadmap for teaching their students. Usually, this document is in the form of a textbook or some other similar compilation. Such a document forms the centerpiece for teacher planning and student instruction.

For too long now education has sought to impart mathematical knowledge to students without placing explicit emphasis on equipping them with the tools and dispositions to assimilate, appropriate, and apply new knowledge and build on it to expand such knowledge and create insights. Because deep understanding is never produced without cognitive processing, engaging students in higher order thinking ought to be an ongoing and central activity in the mathematics education classroom. The primary goal for the Rigorous Mathematical Thinking (RMT) approach presented in this book is to equip students with the dispositions and tools to rigorously engage in systematic mathematics and science conceptual formation with deep understanding. The RMT approach to concept formation involves both the construction of cognitive processes and the utilization of these processes for conceptual development. This approach builds and shapes a networking through a scaffolding mechanism that is guided and organized through mathematical psychological tools to form conceptual structures. A question to be raised at this point is to what extent the standards movement responds to this RMT objective and what has been the standards' impact on helping students function as efficient professionals and ordinary citizens in the United States.

The RMT approach reveals a number of problems stemming from the typical educational practice. First, the standards-based curriculum remains product rather than process oriented. The curriculum states some topic, reviews previous work, provides a ready-made definition of the main mathematics concept, and lays out an algorithm for generating a computational procedure for the symbolic representation of the concept. It also provides an example that walks the students through step-by-step calculations, gives problems to be solved that are on the same level of difficulty as the sample, often showing a few solutions, and then gives homework. In most cases the teacher makes lesson plans based on this routine, attempts to teach lessons, and hopes that students are learning. One of the crucial deficiencies here is the neglect of developing mathematically specific learning activities. Although teachers and students are engaged in many different classroom actions, they are far from the standards of mathematical activity as described in Chapter 1. The construction of mathematical knowledge is absent because the requirements for theoretical conceptual formation and activation and utilization of cognitive processes do not exist. From the RMT perspective it is important that the teacher organizes the material into interconnected conceptual units that lend themselves to systemic development of mathematical knowledge. This organization must include ascending levels of challenge to build intrinsic motivation in students as they progress through the lessons.

A second problem is that the materials and the teacher become the central agents of the teaching/learning process. This contradicts the intent of the standard's movement, which was to make the student the central agent for constructing his or her knowledge. It is not enough for a student to be able to acquire a set of operations and apply them to typical mathematics problems. The students should be capable of learning how to learn and must acquire a rich metacognitive repertoire to allow them to approach new and nonstandard problems and link their everyday experience to the mathematical and scientific knowledge.

A third problem is that students are not given the opportunity to reflect, collaborate, explore, and struggle with ideas. This is needed for the students to develop a personal sense of appreciation of mathematics and experience its intrinsic beauty.

In spite of all of the good intentions of the standards movement, the current approach to teaching science and mathematics concepts in the U.S. classrooms involves the presenting and eliciting of ready-made definitions with accompanying activities that, at best, produce little understanding and superficial applications. The focus in the applications usually does not extend beyond the mechanics or algorithms required for producing concrete answers.

The Need for Rigorous Thought in Mathematics and Science Education

From the mid-1980s to the late 1990s mathematics educators have expressed concern about students' lack of problem-solving skills, reasoning, thinking, and conceptual understanding in mathematics (Hiebert and Carpenter, 1992; Lindquist, 1989). Students are not rigorously engaged in developing and manipulating the deeper structures of their thinking, nor are they challenged to synthesize from their own experiences and knowledge base the understanding necessary to induce the abstractions and generalizations that underlie science and mathematics concepts. Thus, many students complete courses in science and mathematics with the illusion of competency based on memory regurgitation. They do not build the understanding nor the flexible structures required for genuine transfer of learning and the creation of new knowledge in various contexts and situations. These surface experiences are not meaningful to students, do not promote science and mathematics competencies, and to some extent contribute to higher dropout rates. To better understand stronger and weaker aspects of the standards movement it is instructive to look at the difference between American and other systems of education.

Xie (2002) compared the cultivation of problem solving and reasoning in standards in the United States issued by the National Council of Teachers of Mathematics (NCTM) and Chinese national standards issued by the Ministry of Education (MOE):

> Both NCTM and MOE consider problem-solving ability as the main goal of mathematics education. Both of them believe that mathematical problem-solving ability should include both intellectual and non-intellectual aspects. The intellectual aspects include the following contents: the ability to formulate, pose and investigate mathematics problems; the ability to collect, organize and analyze problems from a mathematical perspective; the ability to seek proper strategies; the ability to apply learned knowledge and skills, and the ability to reflect and monitor mathematical thinking processes. The non-intellectual aspect includes the cultivation of positive dispositions, such as persistence, curiosity and confidence, the understanding of the role of mathematics in reality, and the tendency to explore new knowledge from mathematics perspectives. Both NCTM and MOE view reasoning as a process of conjecture, explanation and justification. And both of them believe that mathematics education should foster inductive and deductive reasoning.

There are, however, major differences between NCTM and MOE. MOE advocates the development of cognitive processes that are geared toward conceptual understanding as a problem-solving approach to produce mathematical knowledge and the mindset for mathematical investigation. NCTM, however, supports trial-and-error as a key approach to problem solving that leads to mathematics learning. NCTM views mathematical reasoning as hinging on assumptions and rules through the exploration of conjectures. Again, NCTM promotes the use of trial-and-error strategies along with conjecture strategies and their analyses as means of developing mathematical reasoning. MOE emphasizes developing students' thinking ability as the centerpiece of mathematical ability. MOE's definition of mathematical thought includes mental actions such as comparing, observing, investigating, analyzing, generalizing, abstracting, and reasoning by using induction, deduction, and analogy. Although NCTM presents mathematical language and mathematical thinking as tools for promoting students' abilities in problem solving and reasoning, the details for delineating this development are absent.

Henningsen and Stein (1997) published extensive research on classroom-based factors that support and inhibit high-level mathematical thinking and reasoning. They described the need for students to actively engage in rich, worthwhile mathematical activity to build their capacities to "do mathematics." Their view is similar to that of NCTM (National Council of Teachers of Mathematics, 1991), who believe that mathematical tasks in and of themselves contain the attributes needed for mathematical learning and when students engage in the details of such tasks they gain a sense of what mathematics learning is about. The characteristics of these tasks seem to be of higher levels of complexity than ordinary activity, require greater amounts of time for their successful completion, and involve problem solving, reasoning, thinking, and the capacity to develop mathematical disposition. It is their contention that "the nature of tasks can potentially influence and structure the way students think and can serve to limit or to broaden their views of the subject matter with which they are engaged" (Henningsen and Stein, 1997, p. 525). Although the authors mention mathematical activity, use the terms *generative* and *tools*, and link them to cognitive development in students, their concepts do not fully embody the notion of learning activity (presented in Chapter 1 and further elaborated on in Chapter 3) nor do they fully meet the criteria of mathematical learning activity specified in RMT. There is no mention of general psychological tools, mathematically specific psychological tools, zone of proximal development, or a delineation of students' everyday, spontaneous concepts versus "scientific concepts." Apparently, it is their assumption that student cognition does not require explicit development through the appropriation and utilization of psychological tools.

Reform efforts for mathematics education in the United States are greatly affected by the major beliefs of the stakeholders about the nature of mathematics and its instructional delivery methods and how these will shape classroom practice. The prevailing culture in the United States places a high premium on students' natural ability to do mathematics rather than on student effort. Newton (2007) described in some detail the negative impact such belief has on student motivation and academic performance. He reports that it is common to hear a student say, "I just can't do math." Ability is considered to be a permanent and unchangeable trait rather than a developing function.

This belief categorizes low-ability or high-ability students as equipped for success or failure, respectively, and directly impacts on their level of motivation. Instead, emphasis should be placed on effort, which is an important quality that connects with the idea of rigor in RMT. Effective teaching in the mathematics education classroom should feature an ongoing nurturing and cultivating of effort so that a student will see his or her "intellectual self" in action, making a personal investment in mathematics learning. As such teaching takes place the student will gradually acquire a disposition for perseverance and persistence and develop intrinsic motivation.

Understanding mathematical knowledge is intrinsically conceptual in nature, because mathematics itself consists of "scientific" concepts as described in Vygotsky's (1986) sociocultural theory. A barometer to determine the depth of conceptual learning in the mathematics education classroom is the emphasis placed on the *what, how,* and *why* of learning activity. In U.S. mathematics classes, where the main thrust is placed on the correctness of answers, the dominant focus is on the *what,* with only a little emphasis placed on the *how* and none is placed on the *why* (Chazan, 2000; Hiebert et al., 2005). Greater ability in computational skills and depth in conceptual understanding takes place when teachers design mathematical learning activity that places much emphasis on the *how* and *why* of problem solving and skills development.

Ma (1999) suggests a link among teachers' understanding of mathematical knowledge, their quality of teaching, and perhaps their students' academic performance in mathematics. She states that while most Chinese teachers receive 11 to 12 years of formal schooling, most American teachers receive 16 to 18 years of formal schooling. However, she presents data revealing that more Chinese elementary mathematics teachers demonstrated a profound understanding of fundamental mathematics than their counterparts in America. The paradox continues when we consider that in international studies on mathematics Chinese students outperform American students. Teachers' specialization in mathematics is not a unique "Chinese phenomenon." In such technologically developed countries as Belgium, Japan, and Australia about 50% of primary school teachers have a major or specialization in mathematics, whereas in the United States it is only 28% (Ginsburg, Cooke, Leinwand, Noell, and Pollock, 2005).

At this point it is interesting to elaborate on Ma's concept of what she means by "a profound understanding of fundamental mathematics." From the perspective of the teacher she stresses that mathematical understanding focuses on those qualities of knowledge that facilitate the teacher's capacity to guide students in understanding important mathematical ideas. She delineates four attributes of understanding: (1) basic ideas, (2) connectedness, (3) multiple representations, and (4) longitudinal coherence. She considered basic ideas as those concepts and principles of elementary mathematics that are foundational to all of mathematics. This might be considered as the formative substance of mathematical knowledge on which all other aspects of mathematics are built. When doing more advanced mathematics these ideas are present and must be reinforced. In the RMT approach the teacher must first select and then describe and define the "big mathematical idea" that is to be taught and then perform a structural analysis of this idea in terms of the conceptual components, procedural components, and language of mathematics needed to produce a deep understanding and application mastery of this big idea. The teacher must then perform an operational analysis of this structure, arranging the structural components into a hierarchy from the most fundamental conceptual components, with their required procedural elements, to the more advanced to the original big idea. The big idea, according to Vygotsky (1986), is a "scientific concept" in that it is structural, systematic, and generative and thus presents theoretical models and schemata, psychological tools, and cognitive processes for student learning that are independent of immediately given empirical reality. Though considerably different in scope and definitions, the notion of big idea is similar to Ma's notion of a knowledge package. She considers a knowledge package to consist of conceptual topics and procedural topics that are interwoven to comprise a body of mathematical knowledge understood by the teacher. It is the teacher's understanding of this integrated body of knowledge that equips the teacher with both the information and insight for promoting student learning of the knowledge. It is important that the teacher understands the whole body of mathematical knowledge, the interrelatedness of the pieces of knowledge, and how they form an organized system for mathematical learning.

With regard to connectedness, Ma speaks of the teacher's plan to help students see the relationships among all aspects of the topic of knowledge, ranging from the most superficial and basic to the most complex of concepts, procedures, and operations, thus presenting the learning as a coherent unit rather than a stand-alone, unmeaningful subject. She describes multiple perspectives as the teacher's intention to help students examine an idea from different perspectives often employing different forms of representation. Longitudinal coherence refers to the teacher's knowledge of the entire elementary mathematics curriculum that equips him or her with the understanding and flexibility to link students' learning to previous requirements in lower grades as well as thoroughly and intentionally provide the foundation for what has to be learned in the future at a more advanced level. These qualities of a profound understanding of fundamental mathematical knowledge positions the teacher with the essence of mathematical knowledge and intrinsic effective strategies for leading his or her students to participate with eagerness and personal commitment in acquiring this knowledge.

Ma views the field of elementary mathematics as having depth, breadth, and thoroughness, which demands that teachers' understanding of this knowledge must possess the same qualities for their effective teaching of students. This stipulates that the teacher must aim at teaching his or her students to think to build mathematical knowledge and thus must be metacognitive herself.

The requirements by Ma that teachers have a profound understanding of fundamental mathematical knowledge and that such is essential to effective teaching and learning in students in a way parallels some of the central themes of RMT. The first is that no piece of mathematical knowledge exists in a vacuum but exists in an organized network of ideas and processes. The second is that learning mathematics requires cognitive processes that are mathematically domain specific. RMT has identified and defined cognitive functions or specific thinking actions that construct and manipulate specific mathematics content. The third is that mathematical learning including procedures, operations, and content must be conceptual in nature. There are three major differences, however, between the two. First, it appears that Ma sees mathematical thinking as a process resulting from understanding mathematical knowledge. RMT requires the appropriation and internalization of general psychological tools and the use of these tools to perform systems of cognitive tasks to construct systems of mathematical thinking. Second, RMT views some traditional aspects of mathematics content as mathematically specific psychological tools, such as mathematical symbols; a number system with its place values, number line, table, *x-y* coordinate system, equations and formulae; and the language of mathematics. Such domain-specific tools must organize and orchestrate the use of mathematical systems of thought to construct mathematical knowledge.

The need for rigorous thinking is clearly revealed in a study by Stigler and Hiebert (1997) of 8th-grade mathematics lessons in Germany, Japan, and the United States as part of the Third International Mathematics and Science Study (TIMSS). TIMSS data show that U.S. 8th-grade students scored below their peers from 27 nations in mathematics and below their peers from 16 nations in science. Japanese students scored well above German and U.S. students, whereas German students moderately outperformed U.S. students. Data from the 1996 National Assessment of Educational Progress (Reese, Miller, Mazzeo, and Dossey, 1997) indicated that one-third of U.S. students in grades 4, 8, and 12 performed at the "Basic" level (the performance levels were "Below Basic," "Basic," "Proficient," and "Advanced"). The average international level, however, is also far from adequate. It is telling that a simple TIMSS task of finding out the speed of a car when a graph of the functional relationship between the distance and time is given was correctly solved by only 54.3% of the 8th-grade students internationally (Smith, Martin, Mullis, and Kelly, 2000). These and other research findings point to two gaps in students' mathematics and science academic achievement: overall U.S. students perform below students from some other nations and students internationally perform well below expectations, particularly with regard to conceptual mastery.

It is instructive that the lack of success of U.S. students in the TIMSS does not stem from the lack of exposure to relevant mathematical information. In their study of mathematics performance in the 2003 TIMSS, Ginsburg et al. (2005) demonstrated that the U.S. curriculum omits only 17% of the TIMSS topics through grade 4 and 2% through grade 8 compared to 40% and 25% noncoverage rates, respectively, in other countries. The curriculum of high-achieving Hong Kong omits 48% of the TIMSS items through grade 4 and 18 percent through grade 8. Students in other countries apparently acquire some core mathematical problem-solving strategies that allow them to solve successfully at least some of the problems from the areas that were not covered in their curriculum. These data further strengthen our claim that the key to success in mathematical problem solving crucially depends on the development of cognitive strategies that work beyond the specific operations or techniques.

In the United States a third gap is the performance of minority students versus that of white students. The African American/white and Latino/white academic achievement gaps in mathematics in the United States widened in the 1990s after African American and Latino students' performance improved dramatically during the 1970s and 1980s. Nearing the close of the school year in 1999, 1 in 100 African American students and 1 in 30 Latino students could efficiently do multistep problem solving and elementary algebra as contrasted to 1 in 10 white students. In addition, only 3 in 10 African American and 4 in 10 Latino 17-year-olds had shown facility and understanding in the usage and computation of fractions, frequently used percentages, and arithmetic means contrasted to 7 in 10 white students (Haycock, 2001). This seems to directly affect the advancement of minority students in technological fields. Although minorities account today for about 25% of the American workforce and 30% of the college-age population, they represent just 10% of the bachelor degrees earned in engineering and 6% of employed engineers (Campbell, 1999/2000).

The NCLB Act has complicated the challenges of educational reform more dramatically than any other policy or reform movement in the history of public education. Within the first 4 years of this legislation, criticisms of its accountability mandates, its funding policy to states, and its implementation strategies have reached new heights. Criticism of the NCLB Act transcends socioeconomic class status, ethnicity, gender, religion, and careers.

The two major issues at the forefront of the NCLB debate are accountability and funding. First, the NCLB accountability system requires measures of high-stakes testing, "highly qualified" status for all teachers and paraprofessionals, and local schools meeting the annual yearly requirement of student attendance and achievement. Accountability is enforced at the state level by holding schools responsible for failure to meet or exceed the set expectations. Consequences for failing to meet expectations range from placing a school on probation to shutting it down and restructuring it.

The second issue, education funding, has been shown to be more problematic than the architects of this policy imagined. The NCLB policy places great responsibility on states to monitor and maintain its mandates (Sunderman and Orfield, 2006). States must create and implement curriculum standards and develop the standards-based assessments for reading/language arts, mathematics, and science (added in 2007) that are administered to selected elementary and secondary school grades yearly. In addition, states must assess students with disabilities and English language learners, according to specifications of their respective existing policy guidelines. States must also have a monitoring process in place to ensure that teachers and paraprofessionals are highly qualified. Finally, states must develop data collection and reporting systems and enforce the consequences for local school districts that fail to meet the required expectations. NCLB established a timeline for states to have all of their assessment measures in place and a deadline of 2014 for all teachers and paraprofessionals to be "highly qualified" and students to be "proficient" on state tests, evidenced by disaggregated data. Federal funding for state support of the development, implementation, monitoring, and evaluation of these NCLB mandates was provided during 2002, the first year of implementation, but there was an overall decrease in the following years.

The additional pressure placed on states and local schools translated into pressure for classroom teachers to show dramatic progress in student learning, especially in the areas of mathematics and science. A previous reform effort in mathematics, science, and technology (MST), the Systemic Initiatives (1994–2002), developed and sponsored by the National Science Foundation (NSF), spearheaded the effort to provide high-quality, standards-based, and rigorous professional development and instructional materials for all K-12 MST teachers. The NSF Systemic Initiatives was a precursor to NCLB and helped to lay the groundwork for standards-based MST professional development and accountability for student achievement in mathematics and science.

For the standards movement to succeed three critical needs should be addressed. First and foremost, U.S. students, and indeed all students, must develop the capability and drive to do rigorous higher order mathematical and scientific thinking. Second, elementary and secondary school students must develop a deep understanding of big ideas in mathematics and science and be able to apply them across various disciplines and in everyday living. Third, students must be able to communicate and express their mathematical and scientific thinking orally and in writing with precision and accuracy. It is imperative that the U.S. mathematics and science education enterprise make serious, substantial, and sustained investments in addressing these needs for real academic achievements and transfer of learning to take place for all students. The following two chapters will outline how Vygotsky's sociocultural theory and Feuerstein's notion of mediated learning create the necessary theoretical basis for developing a system of rigorous mathematical thinking that may answer some of the questions posed by the standards movement.

3 Vygotsky's Sociocultural Theory and Mathematics Learning

Mediated Character of Human Learning

For a long time the predominant model of school learning was that of direct acquisition (see Sfard, 1998). Children were perceived as "containers" that must be filled with knowledge and skills. The major disagreement among educators was only in the degree of activity expected of the child. More traditional approaches portrayed the child as a rather passive recipient of prepackaged knowledge provided by teachers, whereas Piagetians and other "constructivists" expected children to be independent agents of knowledge acquisition. In a time it became clear that the acquisition model is insufficient both theoretically and empirically. On the one hand, children proved to be much more than passive recipients of information; on the other hand, students' independent acquisition often led to the entrenchment of immature concepts and "misconceptions" as well as a neglect of important academic skills. A search for an alternative learning model brought to the fore such concepts as mediation, scaffolding, apprenticeship, and design of learning activities.

Vygotsky's (1986, 1998) theory stipulates that the development of the child's higher mental processes depends on the presence of mediating agents in the child's interaction with the environment. Vygotsky himself primarily emphasized symbolic tools-mediators appropriated by children in the context of particular sociocultural activities, the most important of which he considered to be formal education. Russian students of Vygotsky researched two additional types of mediation – mediation through another human being and mediation in a form of organized learning activity (see Kozulin, Gindis, Ageyev, and Miller, 2003). Thus the acquisition model became transformed into a mediation model. Some mediational concepts such as scaffolding (see Wood, 1999) or apprenticeship (Rogoff, 1990) appeared as a result of direct assimilation of Vygotsky's ideas, whereas others like Feuerstein's (1990) mediated learning experience have been developed independently and only later became coordinated with the sociocultural theory (Kozulin, 1998a).

That learning is much more than a direct acquisition of knowledge can be gleaned from the following simple example. Just imagine how a young child learns about dangerous (e.g., hot) objects. According to the direct acquisition model such learning must include a number of direct exposures of the child to a dangerous stimulus, which in due time will result in acquisition of sufficient knowledge that will guide the child's behavior. We know however that aside from cases of severe social-cultural deprivation the human child does not learn about harmful stimuli through direct exposure. They also do not learn about this through the lectures given to them by the caretakers. Instead, a complex process of mediated learning takes place, in which the parents or other caretakers insert themselves "between" the stimuli and the child.

The caretaker indicates to a child which objects are dangerous. Sometimes the caretaker deliberately exposes a child to a dangerous or unpleasant stimulus under controlled conditions, creating the equivalent of psychological "vaccination." The caretaker explains to the child the meaning of dangerous situations. Finally, the caretaker stimulates generalization, creating in the child the notions of a dangerous situation and his possible response to it (Kozulin, 1998a, p. 60).

Even this simple case of mediated learning includes all three major forms of mediation: mediation through another human being, mediation via symbolic tools, and mediation through specially designed sociocultural activity. Caretakers change their behavior to indicate and explain the meaning of dangerous objects to a child in the way attuned to the child's abilities and perspective. Caretakers may focus a child's attention on certain symbolic mediators (e.g., red indicator of a stove or a kettle that becomes for a child a sign of a potential danger). Finally, caretakers may organize a special sociocultural activity, such as a role-play with puppets, the aim of which is to teach children how to avoid dangerous objects.

In a somewhat similar way classroom learning may include all three mediational aspects mentioned above. (1) Acquisition of symbolic tools and their internalization in the form of inner psychological tools then become one of the primary goals of education. (2) Classroom learning becomes organized around specially designed learning activities playing the role of a mediator between students and the curriculum. Instead of just relying on students' generic learning skills such learning activities actively promote students' cognitive development. (3) The role of teacher also changes from that of a provider of information and rules to that of a source of students' mediated learning

Figure 3.1. Pieter Breugel the Elder, *Temperance*, 1560. Klein, A. H., *Graphic worlds of Pieter Breugel the Elder*. New York: Dover, 1963, p. 245. Used with permission of Dover Publications.

experience. This last point was has been elaborated on in considerable detail in Feuerstein's theory of mediated learning (see Chapter 4).

Symbolic Tools and Their Internalization

Let us look at Peter Breugel the Elder's 1560 print *Temperance* (Figure 3.1). Whatever the original intention of the artist, for us this picture provides a rich catalogue of symbolic tools that in the past 500 years has become an integral part of world culture. At the center of the picture stands the allegorical figure of Temperance with a mechanical clock on her head. This is the first of symbolic devices that significantly changed the manner in which people not only do things but also think about them. The earlier time measuring devices, such as a sundial or hourglass, were more like material tools closely connected to natural phenomena such as the movement of the sun and the flow of sand. The mechanical clock separated the material and the symbolic aspects of the tool. The mechanical part became hidden, whereas the symbolic part – the face of the clock with hour signs – became prominently "addressed" to the people.

The ubiquitous presence of the clock in the urban landscape changed the way people related to time; instead of a natural reference to sunrise and sunset the more abstractive reference to such units as hours and minutes began to shape human thought about various phenomena. In a sense, the clock became internalized as an inner cognitive function of thinking whereby units of time are precisely measurable and comparable.

In the lower right corner of the picture we see another "device" that profoundly changed human cognition – the printed book as a tool of learning. The advance of the printing press made books into a rather cheap instrument of school learning. The readily available texts turn educational process from apprenticeship and experience to the analysis and comparison of texts. The second reality – the reality of texts – emerges in addition to the reality of things. Eventually we start thinking textually rather than experientially. This difference is clearly observed when we compare the experience-based and literacy-based approached to problem solving. One of the first studies in this field was conducted by Vygotsky's follower and colleague, Alexander Luria (1976), in Soviet Central Asia in the early 1930s. Luria discovered that the same oral problem is interpreted differently by local peasants who had access to formal education and those who had not. The problems presented to them were of the following type. "There are no camels in Germany. The city of Bremen is in Germany. Are there camels in Bremen or not?" Educated peasants had no problem in accepting this problem (it does not matter whether their answers were correct). Peasants who had no access to formal education refused to accept the problem claiming that because they have never been to Germany they cannot answer the question. When Luria persisted and pointed out that they should pay attention to the words of the question, the peasants responded that "Probably there are. Since there are large cities, there should be camels" (Luria, 1976, p. 112). Formally educated peasants perceived the question as text based with the logic of its own, whereas peasants who received no formal education perceived the same question as related to their own experience or lack of it.

In the lower left corner of Breugel's picture merchants are counting money and entering results into a ledger. This apparently simple device – a table for double-entry bookkeeping – apparently changed the way people from the 15th century thought about their business transactions. Before the appearance of this table merchants recorded their sales and purchases as a continuous text, if you wish, as a story of their work with different goods, places, currencies, expenses, and incomes. The appearance of such a tool as the double-entry table "tamed" this multitude of objects and events and organized thought about them into a uniform pattern. (For more about the influence of this and other symbolic devices on the development of scientific and economic thought, see Crosby, 1997, whose discussion of Breugel's *Temperance* attracted our attention to this picture in the first place.)

Breugel's picture can thus be used as a didactic device demonstrating how a range of symbolic devices – texts, numbers, tables, and musical notations – influenced the way things became perceived and thought about in the culture based on literacy and numeracy. Apart from the fascinating task of reconstructing this cognitive-historical process, there is, however, a much more practical task – how to transfer the required system of symbolic tools to a new generation of learners and how to help them internalize them as their inner psychological tools. This, according to Vygotsky (1979), is one of the primary goals of educational psychology.

In Vygotsky's sociocultural theory cognitive development and learning are operationalized through the notion of psychological tools. Psychological tools first appear as external symbolic tools available in a given culture. Among the most ancient of these symbolic mediators Vygotsky (1978, p. 127) mentioned "casting lots, tying knots, and counting fingers." Tying knots exemplified the introduction of an elementary mnemonic device to ensure the retrieval of information from the memory. A physical object, knot is assigned a symbolic function that then helps individuals to organize their memorization and retrieval. Similarly, finger counting demonstrates how parts of the body (fingers) can serve as an external symbolic tool that organizes cognitive functions involved in elementary arithmetic operations. Casting lots appears in a situation when the "natural" decision is impossible, for example, two alternatives are equally attractive or equally unappealing. This situation is resolved by an application of the artificial and arbitrary tool – die. The individual links his or her decision to the "answer" given by a die, thus resolving the situation that cannot be solved in a natural way. Cultural-historical development of humankind created a wide range of higher order symbolic tools, including different signs, symbols, writing, formulae, and graphic organizers. Individual cognitive development and the progress in learning depend, according to Vygotsky (1979), on the student's mastery of symbolic mediators and their appropriation and internalization in the form of inner psychological tools.

Mathematical education finds itself in a more difficult position vis-à-vis symbolic tools than other disciplines. On the one hand, the language of mathematical expressions and operations offers probably the greatest collection of potential psychological tools. On the other hand, because in mathematics everything is based on special symbolic language it is difficult for a student, and often also for a teacher, to distinguish between mathematical content and mathematical tools. It is easier to grasp this difference in other curricular areas, such as physics, where students have their intuitive empirical experience of certain phenomena and where symbolic tools appear as organizers of this experience. For this reason in what follows we will first provide some examples of psychological tools from curricular areas other than mathematics and only then will turn to mathematical learning itself.

The first stage in this process is the mastery of external symbolic tools and their use as mediators in learning and problem-solving situations. Already at this stage certain students may be at a disadvantage because their home environment does not support such common symbolic tools as pictures, plans, and maps. Culturally different students find themselves in a particularly difficult situation if their native culture does not have some of the symbolic tools routinely used in formal education (see Kozulin, 1998a, Chapter 5). At the same time younger students who successfully master some of the symbolic tools demonstrate the level of reasoning much higher than is usually expected at their age. This phenomenon can be illustrated by the following study conducted by Schoultz, Saljo, and Wyndhamn (2001). The authors took as a point of departure a popular claim that younger children experience serious difficulties in conceptualizing the Earth as a sphere and in reconciling their everyday experience of a flat Earth surface with the idea of a round Earth. Children in Europe have a problem explaining how people do not fall down from the Earth in the southern hemisphere or where a person will find him- or herself after walking for a long time in the same direction. One of the popular explanations is that younger children start with "alternative" concepts of the dual Earth (flat *and* round), hollow sphere, or a flattened sphere. Only by the end of the primary school did students arrive at the concept of the Earth as a sphere. To test this explanation Schoultz, Saljo, and Wyndhamn (2001, p. 110) used the same type of questions as other researchers but introduced a globe as an external symbolic tool:

> To counteract the need of interviewees to orientate themselves in an abstract, verbal framework only, the discussion in our case was carried out using a globe as a point of departure. The globe was placed in front of the child and the interviewer, and the initial question was if the child knew what the object is.

Several interesting consequences were observed. First, even 7-year-old children confirmed both their acquaintance with the globe and the affinity of the features of the Earth with the features of the globe. Moreover, none of the children had any problem with the fact that people live in the southern

Table 3.1. "Stars" task presented to three groups of 7th-grade students ($N = 82$)

Stars

Astronomers classify stars according to their color and brightness. They are also capable of measuring the distance from Earth to each one of the stars and the temperature on their surface. In the following table you can see these data regarding some of the stars.

Stars	Brightness	Distance from Earth (light years)	Surface temperature (C°)	Color
Sirius	1	8.8	10,000	Blue
Knopus	2	98.0	10,000	Blue
Arthur	3	36.0	4,000	Red
Vega	4	62.0	10,000	Blue
Aldebaran	5	52.0	4,000	Red

According to the given data, which two of the properties of the stars are most closely correlated?

1 Brightness and color

2 Brightness and distance from Earth

3 Distance from Earth and color

4 Brightness and surface temperature

5 Color and surface temperature

hemisphere without falling off the Earth. From 84% of the 7-year-olds to 89% of 11-year-olds considered the Earth as a sphere. These results are very different from those typically received in the "shape of Earth" interviews. The authors concluded that:

When considering the outcomes of our study in terms of what they tell us about children's cognitive and communicative capacities, the main conclusion seems to be that when children's reasoning is supported by a cultural artifact such as a globe, they appear to be familiar with highly sophisticated modes of reasoning. (p. 117)

A symbolic tool (a globe) thus proved to be a powerful instrument for shaping young children's reasoning. However, lack of mastery of certain symbolic tools deprives even older students of the success in problem-solving situations that do not require any specific knowledge. For example, in the research conducted by one of the authors (A.K.) it turned out that 13- to 14-year-old students experienced considerable difficulty in solving the "Stars" task (see Table 3.1) that required no specialized astronomy knowledge and can be solved simply by using table as a symbolic tool. If a table were used as a tool for systematic comparison of parameters of different stars the students would easily discover that only two of them, temperature and color, are closely correlated. Nevertheless, only 33% to 40% of the students successfully solved this problem. One may thus conclude that in this case students experienced difficulty at the level of appropriation of a table as an external symbolic tool.

Appropriation, however, constitutes only the first stage that should be followed by internalization of a symbolic tool and its transformation into an inner psychological tool. A table, for example, should become an inner psychological tool that affords learners to think about data in a tabular form. The following research demonstrated, however, that even teachers who have a sufficient mastery of tables as external tools may experience difficulty when asked to spontaneously organize data in a tabular form (Kozulin, 2005b). Teachers who participated in continuing education training were given 24 numbers and were asked to classify them by odd/even and the number of digits (one-, two-, and three-digit numbers) and present the result as a table or chart. Only 48% of them spontaneously selected the optimal form (two columns by three rows) of the table or chart for presenting the data. The rest used nonoptimal tabular forms that failed to provide a proper organization of data and as a result these teachers often "lost" some of the numbers. This result points to the problem of internalization of the external symbolic tool (e.g., a table) and its transformation into an inner psychological tool. Teachers who had no difficulty using tables as external organizing devices showed deficiency in spontaneously thinking about numerical data in a tabular form.

One may classify psychological tools into two large groups. The first is general psychological tools that are used in a wide range of situations and in different disciplinary areas. Different forms of coding, lists, tables, plans, and pictures are examples of such general tools. One of the problems with the acquisition of these tools is that the educational system assumes that they are naturally and spontaneously acquired by children in their everyday life. As a result general symbolic tools, such as tables or diagrams, appear in the context of teaching a particular curricular material and teachers rarely distinguish between difficulties caused by the students' lack of content knowledge and difficulties that originate in the students' poor mastery of symbolic tools themselves. The lack of symbolic tools becomes apparent only in special cases, such as a case of those immigrant students who come to middle school without prior educational experience. For these students a table is in no way a natural tool of their thought, because nothing in their previous experience is associated with this artifact (Kozulin, 1998a, Chapter 5). It would be incorrect to assume, however, that students with a standard educational background spontaneously appropriate and internalize psychological tools. For many underachieving students only a special cognitive intervention built around symbolic tools leads to their acquisition. Instrumental Enrichment (IE; Feuerstein, Rand, Hoffman, and Miller, 1980) is one of the rare cognitive education programs that systematically teach students how to use a variety of symbolic tools – lists, tables, diagrams, plans, maps, graphs – in general problem-solving situations. It will be shown in Chapter 4 how Instrumental Enrichment can be used for creating in the students the necessary cognitive prerequisites for becoming involved in rigorous mathematical thinking.

The acquisition of general psychological tools is a necessary, but definitely not sufficient, prerequisite of rigorous mathematical thinking. What is needed is the acquisition and internalization of domain-specific symbolic tools, in our case the tools associated with mathematical actions. One example of such tools is a number line. A common way of using it in math curriculum is as a representation of a sequence of numbers. However, the notion of representation carries with it a danger of passive acceptance rather than an active use. The number line, as well as other mathematical "representations" should be taught as a tool, that is, as an active instrument allowing students to perform analysis, planning, and reflection. As mentioned above the acquisition of the number line as an external symbolic tool should be followed by the internalization process that ensures that students form a corresponding inner psychological tool. This inner psychological tool will help them to work with number sequences and form precise relationships between quantities or values in the inner mental plane without necessarily referring to an external graphic image. The detailed discussion of mathematically specific psychological tools is provided in Chapter 5.

Zone of Proximal Development

The zone of proximal development (ZPD) is one of the most popular and, at the same time, poorly understood of Vygotsky's (1986) theoretical constructs (see Chaiklin, 2003). One of the reasons why this concept is poorly understood and often misinterpreted is that Vygotsky used it in three different albeit interrelated contexts. The first context in which the notion of ZPD was used by Vygotsky is the context of his developmental theory. Vygotsky (1998) was struggling with the problem of the emerging psychological functions. His argument ran approximately as follows. Typically a child's development is described in terms of already fully formed psychological functions. Such an approach is oriented toward the past rather than the future of the child because it leaves open the question about emerging psychological functions. At the same time these emerging functions play an extremely important role in supporting the central psychological formation that will characterize the next stage of the child's development. That is why it is essential to find the way for revealing these yet "invisible" functions. This can be achieved by observing children's behavior in the context of joint activity with adults or more advanced peers. During this joint activity children will be able to "imitate" only those actions that are based on the emerging functions that belong to their zone of proximal development. Moreover, efficient learning is achieved only when directed at these emerging functions because instead of just adding new knowledge, such learning becomes a true engine of psychological development simultaneously providing new concepts and skills and shaping those psychological functions that "receive" this content.

From the perspective of math education the developmental version of ZPD calls for the analysis of those emerging psychological functions that provide the prerequisites of rigorous mathematical reasoning. Several questions can be asked here. For example, the emergence of which psychological functions is essential for the successful mathematical reasoning at the child's next developmental period? What type of joint activity is most efficient in revealing and developing these functions in the child's ZPD? What characterizes the students' mathematically relevant ZPD at the primary, middle, and high school periods? These questions are directly related to the issue of the relationship between so-called cognitive education and mathematical education. There are reasons to believe that the students' mathematical failure is often triggered not by the lack of specific mathematical knowledge but by the absence of prerequisite cognitive functions of analysis, planning, and reflection. Cognitive intervention aimed at these emerging functions might be more effective in the long run than a simple drill of math operations that lack the underlying cognitive basis.

The notion of ZPD also appears in the context of Vygotsky's critique of static psychometric tests. Vygotsky argued that static IQ tests reveal only the present functioning of the child but say nothing about his or her learning and developmental potential. Thus the static assessment should be complemented by what today is called dynamic assessment (see Lidz and Gindis, 2003). One should be aware, however, that Vygotsky's call for dynamic assessment (DA) was more paradigmatic than methodological. Vygotsky suggested a number of possible intervention methods that may turn the situation of testing from static to dynamic, such as providing a model, starting a task and asking the child to continue, and asking probing questions. However, he never provided an exact methodology of dynamic assessment procedure. This has been done by his followers and by other researchers who developed their own versions of dynamic assessment not related directly to the notion of ZPD (see Feuerstein, Rand, and Hoffman, 1979).

Though one can distinguish quite a number of different types and versions of dynamic assessment one feature that unites them all is the inclusion of the learning phase in the assessment procedure. There are two major formats of the DA, a "sandwich" format that can be used both for individual and group assessment and a "cake" format that is suitable for individual assessment only (Sternberg and Grigorenko, 2002, p. 27). In a "sandwich" type of assessment there are three phases: pretest, intervention, and posttest. At the pretest students are given a complete standard test without intervention. Then comes the intervention phase, which is the core of the dynamic assessment. During this phase students receive instruction in the cognitive strategies relevant to the task. The length and intensity of learning as well as the degree of standardization of intervention vary from one dynamic assessment approach to another. The posttest phase includes either the repeat of the static pretest or an alternative version of it.

In the "cake" format the test is given item by item. If the item is solved correctly the next item is presented, but if the student made a mistake, the intervention process immediately begins to focus on strategies needed for the correct solution of a given item. The intervention may take the form of a sequence of standardized prompts (Campione and Brown, 1987) or of individualized mediation responding to particular difficulties displayed by the student (Feuerstein, Rand, and Hoffman, 1979). Intervention continues until the student is successful in solving the problem, after which a new problem is presented. Sternberg and Grigorenko (2002) suggest that the number of layers in the "cake," that is, the number of hints and mediations required, varies from student to student, whereas the contents of the "layer" may be the same (in the case of prompts) or different (in the case of individualized mediation).

Dynamic assessment procedures have been used predominantly for the assessment of general cognitive functions rather than those related to specific curricular fields. There are several reasons for this state of affairs. Unlike general cognitive functions that are believed to be "fluid" and amenable for change, the functions associated with curriculum material are usually described as "crystallized" (Carroll, 1993) and resistant to short-term changes. In addition, while cognitive dynamic assessment can rely on the same battery of tests (e.g., Raven Progressive Matrices) for students of different ages, curriculum-based assessment by definition requires a much greater variety of materials reflecting the different knowledge base of participating students. That is why some researchers are skeptical regarding the possibility of developing human-mediated dynamic assessment procedures in curricular areas.

> Even for a single school subject a whole range of subject-specific tasks ought to be devised because the competence for solving algebra problems can be vastly different from that for solving geometrical problems. Who is going to construct the great number of procedures that would be required; who will ultimately apply them? The teachers' primary concern is teaching, not diagnosing. (Guthke and Wingenfeld, 1992, pp. 81–82)

All these difficulties notwithstanding, some progress in the curriculum-based dynamic assessment has been made both in the field of reading comprehension and in mathematics. Cioffi and Carney (1983) developed an individual dynamic assessment procedure for evaluating the reading potential of students whose standard test scores indicated significant delay in some comprehension functions. Kozulin and Garb (2002, 2004) designed and tested a group dynamic assessment procedure for English as a second language. In the field of mathematics so far the trend was toward using computers for allowing dynamic assessment that on the one hand is individualized but on the other can be carried out with an entire group of students. Guthke and Beckmann (2000) developed a computer-mediated program for the assessment of students' solutions of number series. Gerber (2000) used the computer-mediated dynamic assessment of learning-disabled students who performed a multidigit multiplication. Finally, Jacobson and Kozulin (2007) demonstrated that the learning potential of students' proportional reasoning that is directly related to such mathematical subjects as fractions and ratios can be evaluated through human-mediated as well as computer-based dynamic assessment.

The third context of Vygotsky's (1986) application of the notion of ZPD is his distinction between "scientific" and "everyday" concepts. Unlike the majority of psychologists who made no clear distinction between a child's concepts developed through everyday experience and those acquired in school, Vygotsky insisted that they have different origins and structure. Everyday concepts are shaped by children's everyday experience and their interaction with adults in the context of everyday, nonacademic activities. These concepts are empirically rich and functionally adequate in a number of concrete contexts and situations. At the same time, some of the most basic of these concepts (e.g., "the sun rises in the morning") do not correspond to physical reality as it is interpreted in scientific study ("sunrise" as a result of the Earth's rotation). Everyday concepts are also episodic, reflecting specific contexts but remaining unconnected to other phenomena that have the same nature. For example, children may have rich everyday experience with the movement of human crowd and the flow of water in the gorge, but unless they have a scientific concept they would not be able to establish a connection between them. Moreover, they would never even guess that these two phenomena are also related to the lift of the airplane, because all three depend on the same principle of the dynamics of fluid known as Bernoulli's law.

Much earlier than the so-called constructivists (e.g., Glasersfeld, 1995), Vygotsky suggested that children do not come to school with an empty head to be filled with academic concepts. Even young children have notions of quantities, structures, causality, and so on. These notions should be taken into account when the teacher begins to present academic mathematical, physical, or historical concepts in the classroom. At each given moment in the classroom discourse as well as in the students' minds there is a complex interaction between original everyday concepts of the students and academic concepts provided by teachers. In this context ZPD is interpreted as a zone of possible dialogue between academic and everyday concepts. A strong aspect of academic concepts is their systemic nature and the conscious character of their application. Their weak point is their abstractive verbal nature detached from the child's experience. Thus teaching in the ZPD presupposes the application of the system of academic concepts to the phenomena empirically familiar to students. As a result, everyday phenomena cease to be linked to a specific context and become transfigured into the particular instance of a general scientific principle. The zone of a dialogue between academic and everyday concepts is delimited from both sides. It would be counterproductive to introduce scientific concepts that do not find the relevant material in the students' everyday experience. A certain level of the development of everyday concepts is essential. At the same time it would also be wrong to apply a higher order scientific concept that does not have a proper support in the already established scientific notions of the lower level. A systematic, hierarchical nature of scientific concepts provides a proper guideline for establishing progressively changing ZPDs of the students.

Vygotsky, however, was too categorical in associating formal educational settings with the development of rigorous scientific concepts. Vygotsky's followers' (Davydov, 1990) analysis of primary and middle school curricula and instructional methods revealed their considerable reliance on simple empirical generalizations of everyday phenomena. Instead of being radically transformed, children's everyday concepts are often simply enriched and organized. As a result many of the children's "misconceptions" are actually further entrenched by this school practice. Instead of studying in the ZPD students just expand their already existent empirical reasoning to new objects and their properties. What was advocated by Davydov (1990) and other Vygotskians (Karpov, 2003a) as an alternative was to introduce the principles of theoretical rather than empirical learning as early as primary school. Theoretical learning presupposes the analysis of every object or phenomenon in terms of its essential features and creation of its model. This model is then manipulated so as to determine the properties and boundaries of the phenomenon. Students thus acquire a true conceptual understanding of the object or process as well as develop their own cognitive and metacognitive skills. Effective teaching in ZPD thus requires a reorientation of the instructional process from empirical to theoretical learning. This can be achieved through the design of special learning activities discussed in the next section.

Learning Activity

Sociocultural theory makes an important distinction between generic learning and specially designed learning activity (LA). This distinction is made in the context of Vygotskian interpretation of developmental periods as dominated by a particular type of activity (see Karpov, 2003b). Formal learning becomes a dominant form of child's activity only at the primary school age and only in those societies that promote it. Generic learning on the other hand appears at all the developmental ages in the context of play, practical activity, apprenticeship, interpersonal interactions, and so on. In a somewhat tautological way specially designed LA can be defined as such forms of education that turn a child into a self-sufficient and self-regulated learner. In the LA classroom learning ceases to be a mere acquisition of information and rules and becomes learning how to learn (see Kozulin, 1995). Graduates of the LA classroom are capable of approaching any material as a problem and are ready to actively seek means for solving this problem. Needless to say, not every formal educational setting meets the LA requirements. The majority of these settings just use students' generic learning abilities with the aim of providing students with information and skills. That is why the LA approach had to formulate its own methods of instruction and design its own instructional tools.

Some of the basic principles of the LA approach were formulated by Vygotsky (1986) himself, whereas others were elaborated on by his students and followers (see Kozulin, Gindis, Ageyev, and Miller, 2003). This approach places educational process as a source rather than a consequence of the development of a child's cognitive and learning skills. According to Vygotsky's model education does not coincide with development but should be designed in such a way as to promote those psychological functions that will be needed during the next educational phase. In this way the LA approach clearly differs from educational constructivism based on Piagetian theory (Duckworth, 1987). It is true that both approaches emphasize students' activity and the constructive nature of their concept formation. Instead of receiving concepts from the teacher in a ready-made form students are expected to actively construct them. However, whereas Piagetian constructivism perceives students as natural learners who will use their existent cognitive skills to construct the required concepts, the Vygotskian LA approach takes the students' natural abilities only as a starting point. Students' development depends on the educational process guided by the teacher. If in Piagetian constructivism students' cognitive functions are viewed primarily as dependent on maturation and personal experience, in the LA approach they are perceived as products of a deliberately designed educational intervention. Thus whereas Piagetian constructivists take the existent cognitive functions of the students as a basis for the learning process, in Vygotskian LA theory students' cognitive development appears as an outcome of learning and instruction.

Moreover, whereas the majority of educational approaches, both constructivist and more traditional, are based on the dichotomy of cognitive functioning and curricular content, the LA approach overcomes this dichotomy by construing curriculum as a concept formation activity that directly contributes to students' cognitive development. Reading, writing, and math operations are treated in LA theory on equal footing with other higher mental processes. As a result, conceptual knowledge becomes inseparable from the process of concept formation, which in its turn shapes the students' cognition.

Conceptual learning characteristic of an LA classroom is clearly distinct from the situated everyday learning of apprenticeship type (Rogoff, 1990). These two types of learning belong to different sociocultural contexts and different types of activity. In the LA classroom learning is aimed at developing students' systematic "scientific" concepts in all fields of knowledge, not only in natural sciences. The apprenticeship type of learning leads to the development of everyday concepts that are experientially rich and practical in a given context, yet often incompatible with the scientific picture of the world (Karpov, 2003a; Stech, 2007). The systemic nature of LA responds to the systemic character of disciplinary knowledge. As physical, biological, or mathematical knowledge is not a collection of isolated facts but a conceptual structure within which each element is connected to another, in the same way LA develops students' comprehension as a system of conceptual actions rather than a collection of experiential episodes.

The above general premises of LA theory are translated into more specific goals, materials, and methods of classroom instruction. It is important to realize that the LA approach does not aspire to solve all the problems of the developing child. Such qualities as moral integrity, empathy, or spontaneous flight of imagination lie outside the scope of LA and must be promoted by different types of activity (Zuckerman, 2003). The exclusive goal of the LA approach is to develop the child as an independent and critical learner. That is why classroom activities designed according to the LA model are dominated by the objective of learning how to learn, whereas other aspects, such as socialization or acquisition of information, play subdominant roles. Students in the LA classroom are infused with the conviction that learning how to learn is the focal point of their school experience.

Three elements constitute the core of LA: analysis of the task, planning of action, and reflection. Analysis and planning feature prominently in many educational models, at the same time reflection as a central element of the primary school education may justifiably be considered a "trademark" of the LA approach. One may ask what justifies such an early emphasis on reflection. Why not use the first years of schooling for the development of basic skills and then, in middle school, using children's more mature cognitive functions, shift the emphasis on reflection. According to the LA approach the problem is that traditional, nonreflective teaching of basic skills leads to the development in students of certain bad learning habits that are very resistant to correction in middle school (Zuckerman, 2004). Among these habits are working exclusively according to a previously given model, expecting only one correct answer to any task, accepting an authoritative position without examining it, as well as lack of tools for evaluating one's own and others problem-solving strategies. An early start of reflective learning guarantees that these habits do not get entrenched in students' minds.

According to Zuckerman (2004) there are three major aspects of reflection to be developed in primary school: (1) the ability to identify goals of one's own and other people's actions, as well as methods and means for achieving these goals; (2) understanding other people's point of view, involving looking at the objects, processes, and problems from a perspective other than one's own; and (3) the ability to evaluate oneself and identify strong points and shortcomings of one's own performance.

For each one of the above aspects of reflection special forms of learning activity were developed:

- Development of the ability to identify goals, methods, and means of action requires creating a mental schema of the action. Students are encouraged to analyze actions into their constituent parts and use symbolic tools such as signs, symbols, and schematic drawings to represent the action schema.

Eventually these symbolic tools become internalized as inner psychological tools and students start using them spontaneously for symbolic representation of any new action.

- Understanding another person's point of view is developed through cooperative peer learning and by older students teaching younger students (e.g., an 11-year-old teaching a 7-year-old). During these activities students are explicitly instructed to reflect on problem-solving strategies of the other.
- The ability of self-evaluation is promoted by teaching students how to select evaluation criteria and how to build evaluation scales. Through these type of activities students learn that evaluation is not a subjective judgment delivered by a teacher but an objective process depended on selected criteria and standards.

Learning Activity and Math Curriculum

Although three major aspects of reflection are similar in all curricular areas, the instructional methods and materials vary from one area to another. We focus here on the LA math curriculum for the primary school (Davydov, Gorbov, Mikulina, and Saveleva, 1999; Schmittau, 2003, 2004). The LA curricular content and the methods of instruction differ significantly from those used in both traditional ("back to basics") and constructivist classrooms. Probably the best illustration of this difference is the treatment of the notion of "number" and later on the notion of "fraction" in the LA curriculum. First, number does not constitute the starting point of the LA 1st-grade math curriculum and when introduced it is not associated with counting discrete objects. Before numbers are introduced children in the LA classroom are taught how to handle comparison of different quantities, such as length, weight, area, and volume. Initially the quantities are selected in such a way that children can see that one of them is bigger than the other without placing them side by side. Children are then presented with quantities that should be aligned to determine which one is bigger. Already at this stage children are introduced to symbolic representation of equality (=), as well as "bigger" ($>$) and smaller ($<$). Simultaneously they are introduced to the methods of selecting a parameter of comparison (e.g., color, length, and area). Then children are confronted with the task of comparing quantities that cannot be aligned (e.g., the length of the desk and the height of the bookcase). The children are led to the discovery that they need an intermediary (e.g., a piece of rope). In this way they can affirm that the height of the bookcase is indeed bigger because the length of the rope that equals the length of the desk is less than the height of the bookcase. After children master the use of intermediaries for the purpose of comparing different quantities, they are given a task of comparing two line segments by using a short strip of paper as an intermediary. By applying the strip of paper to each one of the segments children arrive at their measure. The measure is defined as a ratio of the length of a segment to the length of the measuring unit. These operations immediately receive symbolic representations. For example, if the first segment is designated as A and the measuring unit as u, then A/u is a measure. On this basis numbers are introduced through the act of measurement, and a number is thus defined as a ratio between a certain quantity and the unit of measurement. The advantage of this approach, which is implemented from the very beginning of classroom learning in the 1st grade, is that instead of strengthening the everyday notion of a number as related to countable objects and thus limiting the students' understanding to only positive integers, it opens for students the possibility of constructing any number, including fractions and even irrational numbers.

Educational research accumulated sufficient evidence that children whose notion of number is based on natural numbers experience serious difficulties when confronted with other types of rational numbers, such as fractions (Vamvakoussi and Vosniadou, 2004). Among conceptual and operational misconceptions engendered by the natural numbers paradigm is a belief that numbers with a greater number of digits are bigger, that multiplication always leads to bigger numbers, and that there is either no or a finite number of numbers between two pseudosuccessive numbers (e.g., "no numbers between 0.006 and 0.007" or "there is only one number, 4/7, between 3/7 and 5/7"). Though there is little doubt in the existence of these misconceptions, their origin can be interpreted differently. One may present the acquisition of natural numbers as a natural process that has a privileged position because it has innate neurobiological or maturational basis (Vamvakoussi and Vosniadou, 2004). From this point of view learners inevitably face a difficult conceptual change because their natural concept of numbers at a certain stage should be replaced by a scientific notion of rational numbers. The LA approach, on the contrary, claims that the main reason for the persistence of students' misconceptions is that these misconceptions are actively supported by a predominant educational approach that introduces numbers through the counting procedure. What is needed, therefore, is not a conceptual change in the older students but a proper conceptual and operational introduction of numbers as a ratio in younger learners.

As with many other LA curricular innovations, the introduction of number as a ratio is important not only for the 1st-grade curriculum within which it is taught but also as a basis for all subsequent mathematical developments. It prominently demonstrates its potential when LA classroom students reach the 4th grade and start learning fractions (Davydov and Tsvetkovich, 1991). Because students are already used to measurement as a way to conceptualize numbers, they experience no significant discomfort when they discover that quantity A equals 2 measures U plus a remainder. Then students find a new smaller measure K, so, for example, that $3\ K = U$. By applying the measure K to the remainder they find that the remainder equals 1 measure K. Returning to the question about A, they conclude that A equals 2 measures U and 1/3. When asked, "One-third of what?" students respond, "one-third of measure U" (Morris, 2000, p. 47). Moreover, $9\frac{1}{2}$-year-old children studying in an LA classroom arrive at the conclusion that between 0 and 1 are "a lot of numbers; too many to count" (Morris, 2000, p. 72), whereas in the regular classrooms only 10% had this opinion (Stafylidou and Vosniadou, 2004). In the LA classroom 5th-grade students are quite capable of handling a problem like "What is the amount for X in the expression $22/39 + X/39 = 2$" (Atahanov, 2000), which is difficult for much older students in regular classrooms.

The content of LA curriculum and its instructional methods necessitated also a different type of the primary school math textbooks and teacher manuals. As one can expect, the issue of reflection features prominently as early as the beginning of the 1st-grade math textbook (Alexandrova, 1998). For example, in the task of finding shapes with the same area children are encouraged to first "teach" their teacher how one should approach this task and only after that actually perform the necessary operations. After accomplishing the task, children are asked to invent their own problems based on the same shapes. The textbook also includes exercises directly aimed at correcting the lack of conservation (or number, area, or volume) that Piaget showed to be quite typical for 6- to 7-year-old children. For example, children are given the homework of finding containers that have the same volume but a different form. Children are expected to explain their way of selecting specific containers. In the classroom, the teacher shows two bottles of a very different form with colored water reaching the same level in both of them. The teacher tells children that she poured the same amount of water in both bottles and asks children to convince her that she is wrong.

Very early children are taught to use all kinds of symbolic representations, for example, letters for designating values and line segments for schemas:

$$D = C\ (D \text{ equals } C);\ P > F\ (P \text{ bigger than } F)$$

Using these formulae build a schema made of line segments. Find two or more objects that fit these formulae and schema.

Schema:

________________ ___________

________________ _____________

Use letters to designate values in this schema; write the formulae.

On this basis children are asked not only to integrate schemas and formulae but also to find "a catch" deliberately included into the tasks. For example, when schema do not correspond to the formula, the child may respond that the formula $A = B$ reflects the height of two bottles, whereas the schema:

____

____

corresponds to their volumes.

One of the very simple but powerful schematic devices is the part/whole schematic "∧":

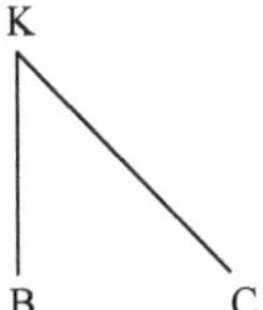

This schematic enables students to see part/whole relationships in the variety of mathematical expressions whose different surface features obscure their common theoretical structure (Schmittau, 2004). For example, this schematic allows us to see that there is a common theoretical structure in the following tasks:

$6 = 4 + 2$; $234 = 2(100) + 3(10) + 4(1)$ and "John has 7 baseball cards. Tom gave him some more and now John has 15. How many cards did Tom give him?"

6

∧

4 2

$7 + X = 15$

15
$\wedge$
7 X
$X = 15 - 7$
$X = 8$

The use of the $\wedge$ schematic as a tool of analysis enables children to see that the missing term X is a part and must, therefore, be obtained by subtracting the known part, 7, from 15, which is the whole. The schematic enables children to distinguish actions on quantities from actions on numbers; the quantities may be added, but the numbers representing their measure or count may need to be subtracted to find a missing measure of count (Schmittau, 2004, pp. 27–28).

Reflective self-evaluation is achieved by including the following questions into the 1st-grade textbook (Alexandrova, 1998, p. 86):

Find the tasks in this chapter that you consider to be:

1) The most important; 2) The most interesting; 3) The easiest ones; 4) The most boring; 5) The most difficult. Explain your choice and compare it to that of other children.

As mentioned earlier three elements constitute the core of learning activity: analysis of the task that discerns its core principle; systematic planning of the problem-solving process; and reflection on one's own presuppositions, actions, and results. These three aspects can also be used for evaluation of the development of students' general and mathematical reasoning (Atahanov, 2000). A considerable number of students of different ages stay at the preanalytic, purely empirical level of problem solving. They take into account only surface features of the tasks and tend to apply the standard algorithm even when it is absolutely inapplicable to a given task. For example, 13- to 14-year-old students at the empirical level solve the following task by just performing division by 8 without any attempt to discover the task's core principle:

$33:8 = ?; 39:8 = ?; 41:8 = ?; 47:8 = ?; 49:8 = ?; 55:8 = ?; 57:8 = ?; 63:8 = ?$

Students who perform at the analytic level will discover the core principle of the above task, but they may still experience difficulty with planning. Such students experience difficulties with tasks like:

Tania is younger than Nadine by 5 years. Zelda is younger than Nadine by 2 years and 6 months. What is the difference in age between the oldest and the youngest girl?

Finally, students who demonstrated both analytic and planning abilities often fail at the tasks that require reflection. For example, the following task requires serious reflection on the terms of the task:

> Write down a general algebraic formula for all numbers that being divided by 5 have a remainder 7. (Atahanov, 2000, p. 197)

At the first glance the task is "strange" because the remainder is bigger than 5. However, if students overcome this feeling of strangeness they can arrive at a rather simple formula that will provide the answer ($N = 5n + 7$; for $n \geq 1$).

Atahanov (2000) analyzed the level of mathematical reasoning in 12- to 17-year-old students studying in regular (non-LA schools). For each grade its own set of tasks responding to analytic, planning, and reflective levels of reasoning was created. It turned out that only 11% to 12.5% of older students reached the reflective level, whereas 65% to 75% remained at the preanalytic, empirical level. At the same time only 31% of the students studying in the 4th grade of the LA classrooms demonstrated empirical reasoning, whereas the remaining 69% solved the tasks at the analytic, planning, and even reflective levels.

Similar results were reported by Zuckerman (2004), who compared the math problem solving of students in two Moscow schools, one an LA school and the other an elite school that used a regular curriculum. Students at both schools were given PISA-2000 (Program for International Student Assessment) math tasks. The tasks differed in their complexity and type; some problems required efficient application of standard algorithms, whereas other required reflective reasoning. With the standard problems the advantage of LA school students was not that great: 78% of the LA students solved the more challenging of the standard problems versus 68% in the traditional school. However, with the reflective problem the gap widened considerably: 71% in the LA school versus 42% in the traditional school. One may conclude that systematic implementation of LA curriculum indeed leads to a much better math performance with the tasks that require analysis, planning, and reflective reasoning.

Another aspect relevant to the issue of LA concerns the relationship between the level of general reasoning and mathematical reasoning. In his research Atahanov (2000) advanced the hypothesis that students who demonstrate a particular general cognitive level of reasoning (analytic, planning, or reflective) will show mathematical problem solving at the level not higher than their general level. For example, students who demonstrated general problem solving on the planning level may show math problem solving on the planning, analytic, or empirical levels but not on the reflective level. The analysis of the general and math problem solving of 155 students confirm Atahanov's hypothesis. For example, although in general problem solving 50.3% of students demonstrated a preanalytic, empirical level of reasoning, in math problem solving this number grew to 78.1% and this was "at the expense" of students who demonstrated an analytic or planning level of reasoning in general problem solving. Atahanov's (2000) results strongly support the RMT paradigm that presupposes two major targets of action, students' general cognitive functions and specific mathematical functions and tools. To reach the reflective level in math problem solving students should first acquire analytic, planning, and reflective general cognitive skills. This can be achieved by implementation of the IE program (see Chapter 4). Tools and strategies acquired through IE create in the students the basis for acquisition of math strategies and tools. One may say that advancement of general cognitive functions creates a ZPD for more special mathematical functions. Thus the task of the teacher is, on the one hand, to advance the student's general cognitive level, thus creating the potential for math learning, and, on the other, to realize this potential in the form of mediation of math-specific tools and strategies.

4 Mediated Learning and Cognitive Functions

Sociocultural theory (see Chapter 3) identified three major classes of mediators interposed between learners and their environment: (1) physical mediators, such as material tools and technologies; (2) symbolic tools, such as signs, languages, and graphic organizers; and (3) human mediators, such as parents, teachers, peers, and other mentors. The learning process, therefore, is rarely immediate. Sociocultural mediators are ubiquitously present in the life of a child, first as a simple tool such as a spoon, with which the child develops motor skills, then as language that becomes a tool of thought, and then as a parent or teacher whose intervention ensures the child's acquisition of material or symbolic actions. From the very beginning children actively interact with the above mediators and with time internalize their actions as their own inner psychological functions.

This perspective, however, differs from that envisioned by Jean Piaget, whose concept of child development was probably the most influential psychological theory in the second half of the 20th century. This is how Piaget (1947/1969, p. 158) describes the interaction between infants and their environment:

> ... Seen from without, the infant is in the midst of a multitude of relations which forerun the signs, values and rules of subsequent social life. But from the point of view of the subject himself, the social environment is not necessarily distinct from the physical environment... People are seen as pictures like all the pictures which constitute reality... The infant reacts to them in the same way as to the objects, namely with gestures that happen to cause them to continue interesting actions, and with various cries, but there is still no exchange of thought, since at this level the child does not know thought; nor consequently, is there any profound modification of intellectual structures by the social life surrounding him.

There is a good chance that a young Israeli student, Reuven Feuerstein, who came to the University of Geneva in the late 1940s, heard this description from Piaget himself. Piaget, however, did not find it necessary to inform his students about the opposite, sociocultural point of view promoted by Vygotskians, with whom he was in contact as early as the late 1920s. Thus Piaget's students were left to their own devices if they wished to depart from the magisterial worldview of their teacher. With the wisdom of hindsight (see Feuerstein, 1990) we can identify two major points where Feuerstein's vision deviated radically from that of Piaget. The first point concerns the role of human mediators in the life of children beginning in early infancy. The second is related to the modifiability of the child's cognitive structures under the influence of interaction with human mediators.

According to Feuerstein (1990) mediation provided by parents and other caregivers constitutes a decisive factor in the child's development from a very early age. He postulated that children's learning occurs in two forms: direct learning based on the immediate interaction between children and their environment and mediated learning that depends on another human being (parent, teacher, etc.) placing him- or herself between the environmental stimuli and the child. A human mediator thus selects, amplifies or reduces, repeats, schedules, and interprets environmental stimuli for the child. Moreover, Feuerstein maintains that an appropriate experience of mediated learning constitutes a prerequisite of efficient direct learning. Mediated learning experiences of children establish the basis for efficient learning and problem-solving strategies they apply to increasingly more difficult tasks throughout their childhood and into adult life:

> It is our contention that mediated learning experience provides the organism with instruments of adaptation and learning in such a way as to enable the individual to use the direct-exposure modality for learning more efficiently and thus become modified... On the other hand, the individual lacking mediated learning experience remains a passive recipient of information and is limited in his capacity for modification, change, and further learning through direct exposure.... (Feuerstein, Krasilovsky, and Rand, 1978, p. 206)

The above model suggests a new interpretation of the relationship among genetic, organic, social, and psychological factors on the one hand and the developmental outcomes on the other. Usually the first set of factors is perceived as a direct determinant of the child's development. Children's development and achievements are linked directly to favorable or unfavorable genetic, organic, and social circumstances. Children with certain genetic syndromes (e.g., Down syndrome), organic impairment (e.g., cerebral palsy), or social conditions (poverty) are expected to be at risk of increased learning and performance problems and decreased developmental outcomes than children with a more favorable set of factors.

Clinical and educational practice, however, provides endless examples of cognitive-developmental outcomes that cannot be explained on the basis of the direct influence of genetic, organic, or environmental influences. One and the same form of cognitive deficiency or learning problems is often associated with quite different sets of organic and environmental factors while apparently similar organic and environmental combinations often lead to significantly different cognitive-developmental outcomes ranging from normal to pathological. Feuerstein's theory (1990) suggests that genetic, organic, and social factors constitute only distal determinants of cognitive development, whereas a mediated learning experience (or the lack of it) constitutes the proximal determinant.

There are, however, intricate dialectic relationships between the proximal factors and mediated learning. On the one hand, mediated learning can moderate the influence of unfavorable organic or environmental factors. On the other hand, these same factors may prevent adequate mediation from taking place, which in turn affects the child's ability to benefit from direct learning. For example, extreme poverty may prevent parents from spending any meaningful time with their children, which reduces the amount of mediated learning and thus negatively affects the necessary prerequisites for direct learning. On the other hand, this negative impact might be significantly alleviated by the extra mediation provided to the children by other members of the extended family (e.g., grandparents). As a result, mediation may both moderate the direct impact of poverty and suffer from this impact.

The second point of Feuerstein's radical departure from the Piagetian model concerns the cognitive modifiability of the child. In Piaget's theory (1947/1969) cognitive change reveals itself in transition from one developmental stage to the next (e.g., from sensory-motor to intuitive intelligence or from the stage of concrete to the stage of formal operations). Maturational processes and a child's direct interactions with the environment are responsible for this universal progression. External factors may prompt the child who is almost ready for the next stage to "cross the border" but they are not capable of radically changing the immature cognitive structures of the child in a short period of time. Feuerstein (1990), on the contrary, asserted that a radical modifiability is possible with the absent cognitive structures literally constructed via mediated interactions. The notion of structure in Feuerstein's theory thus both uses the main ideas of Piagetian structuralism and negates them. Similar to Piaget, Feuerstein asserts that cognition and learning cannot be efficient if they remain on the level of atomized skills, behaviors, or bits of knowledge. Efficiency and flexibility of thought are achieved through the formation of generalized structures responsible for cognitive solutions over a broad range of tasks. Agreeing with the necessity of a structural approach, Feuerstein, however, does not accept the Piagetian explanation regarding the formation of cognitive structures.

To understand the possible sources of this disagreement it might be appropriate to dwell on the differences in the actual child-related experiences that provided very different perspectives for Jean Piaget and his student. Piaget's research focused almost exclusively on middle-class Swiss children who had a stable family environment and standard educational experiences. It is these children – who by and large faced no particular challenges – who served for Piaget as a model for the child in general. Feuerstein's experiences could not be more dramatically different. Feuerstein started his career as an educator and counselor for child survivors of the Holocaust. There were almost no standard features in the development and environment of these children. They often were orphaned or separated from their parents for prolonged periods of time. Their language development was erratic. Many of them had no formal education or just disjointed segments of learning acquired in different languages and under the most nonstandard conditions, such as a concentration camp. For a practical educator the question was thus not about the natural developmental stage of these children – it was pretty clear that they lagged behind the age norm – but how to change their cognitive and learning situation. During the next stage of his career, in the early 1950s, Feuerstein worked with refugee children from North Africa who had their own share of cultural, social, and familial dislocations. Here again his concern was not so much to determine the mental age of the children but to find a way to prepare them for integration into modern classrooms in the new country.

Two conclusions can be drawn from this short excursion into the history of Feuerstein's educational experience. First, his perspective from the very beginning was aimed at inducing a change in the children's cognitive status rather than dispassionately investigating this status. Second, Feuerstein's observations led him to believe that human mediation plays an important role in the formation of cognitive structures or in their remediation.

For our discussion of Rigorous Mathematical Thinking (RMT) the issue of structural cognitive change is relevant in all three of its constituent aspects: structure, cognition, and change. We claim that successful mathematical thinking is impossible without creating cognitive structures in the child's mind, first more general structures required for any type systematic learning and then specific structures of mathematical reasoning. Structures provide both the organization of thinking and its systematicity. Without them child's mathematical thinking would remain a disorganized collection of pieces of information, rules, and skills that does not possess the required generality or rigor. The emphasis on cognition stems from our conviction that a considerable part of students' difficulties in mathematics stems not from the lack of specific mathematical information or procedural knowledge but from the underdevelopment of general cognitive strategies required for any systematic learning. Mathematical knowledge itself would remain latent if not activated by the relevant cognitive processes. Finally, our aim is to generate the change in students' mathematical reasoning rather then just observe its "natural" development or lack thereof. Actually both Vygotsky's and Feuerstein's approaches reveal that such a "natural" development is rather illusory. What is observed as a "natural" development in monolingual middle-class children with a standard learning experience might appear "exceptional" in refugee children who lack formal educational experience and struggle to learn a new language and new rules of the educational game. Thus the question is not what "naturally" happens in the children's mathematical reasoning but how to construct the reasoning that corresponds to a given sociocultural goal.

Mediated Learning Experience

The idea that mediation provided by parents, teachers, and other mentors is beneficial for a child's development is not particularly original. Only the most radical individualists would insist that everything in a child's development comes from his or her genetic endowment and direct learning experiences unaffected by mediation. There is, however, a considerable difference between acknowledging the generally beneficial impact of mentors' mediation and providing a systematic elaboration of what kind of mediation is beneficial for children's development and learning. It is for this reason that one must distinguish between a generic term, *mediation*, and the criteria of a mediated learning experience (MLE) elaborated on by Feuerstein (1990).

Not every situation that involves a child, a mentor, and a task leads to the experience of mediated learning. According to Feuerstein, at least three criteria, intentionality, transcendence, and meaning, should be present to render such an interaction of the quality of MLE. Intentionality of the interaction implies that mentors constantly attune their behavior to the goal of attracting and keeping the child's attention as well as making the task accessible to the child. Moreover, the child is made aware of the deliberate rather than accidental nature of the interaction among mentor, task, and child. The importance of the teacher's intentionality in the classroom can best be described through negative examples. The teacher who lets students sit at the back of the classroom and continue their quiet conversation when the class is engaged in general discussion deprives them of intentionality and thus reduces their chances of gaining MLE. If the presentation of learning material is done formally, simply because "this is how it is written in the textbook," there is a great chance that such an absence of intentionality on the part of teacher will lead to the lack of MLE in students. A good teacher is constantly in search of special techniques for making material accessible to this student or this group of students. The intentionality may be not only absent but also misdirected. For example, some parents, while helping their children with homework, primarily have in mind the effect that this activity has on their spouse rather than on the child. They may actually be quite successful in what constituted their real goal, but for the child this interaction is devoid of MLE. Finally, it is quite typical for college professors giving a lecture to show off their erudition so that students leave the lecture hall with a firm belief in the inadequacy of their own understanding but without a grain of MLE. Intentionality thus is achieved by constantly monitoring students' needs, skillfully sustaining their attention, deploying various techniques for adjusting the learning material to students' perception and activity, and making students aware that learning is not an accidental but thoroughly deliberate process.

The second criterion of MLE is transcendence. This is how Feuerstein (1990, pp. 97–98) introduced it:

> The mediator does not limit the length and breadth of the interaction to those parts of the situation that have originally initiated it. Rather he or she widens the scope of the interaction to areas that are consonant with more remote goals. By way of illustration, if the child points to an orange and asks what it is, a non-mediated answer will be limited to simple labeling of the object in question. A mediated transcendent interaction will offer a categorical classifying definition: 'It is the fruit of a plant, a tree. There are many fruits similar to the orange: a lemon, a mandarin, etc. They are all juicy. Some are sweet, some are sour, some are big, others small. They are all citrus.' In transcending the immediacy of the required interaction, the mediator establishes a way in which the mediatee can relate objects and events to broader systems, categories, and classes.

Mediation of transcendence leads the child beyond the "here and now" situation or task. This is one of the central and, at the same time, most difficult objectives of any educational system because it addresses the following paradox. On the one hand, children should be taught everything that they do not know; on the other hand, no educational system, whatever its scope and intensity, can teach "everything." Thus we must teach only certain things but in such a way that this learning experience can then be applied to the tasks that lie beyond what has been actually taught. As in our discussion of intentionality, the importance of transcendence is best demonstrated through negative examples. If in the mechanics class students are just taught how to select an appropriate formula and put certain numbers into it, this experience would not help them in the study of electricity. Instead of learning the general principles of scientific reasoning as applied to a mechanics problem, they only studied concrete operations of manipulating formulae. When students are shown how to put population data into a table that represents different states of the United States but are not taught about the table as a general tool-organizer, they most probably would not be able to independently select an appropriate table for biological data analysis. On a more general level one may say that when a classroom activity starts and ends with the material presented during this specific lesson, the aspect of transcendence is missing and thus the MLE is absent. What is distinctive in Feuerstein's (1990) approach to the issue of transcendence is its comprehensiveness. His illustrations of transcendence include a very wide range of situations from meal-time behavior of toddlers to traditional family storytelling to specially designed cognitive enrichment activities in the classroom.

Different aspects of transcendence have been discussed by a number of researchers under various names, such as transfer, generalization, and bridging. Thus Perkins and Salomon (1989) convincingly demonstrated that one of the tacit assumptions of many educators is that if you teach specific operations to the children they will somehow spontaneously generalize them and transfer them to a different context and material. This assumption, however, is far from being empirically supported. On the contrary, all the evidence indicates that children do not perform spontaneous transfer if they are not made aware of this and are not taught the strategies of transfer. Moreover, as demonstrated by Brown and Ferrara (1985), there is no unequivocal connection between students' ability of direct learning and their transfer abilities. Some fast learners turned out to be rather poor at transfer, whereas not so fast learners demonstrated considerable transfer abilities. The followers of Vygotsky (see Davydov, 1990; Zuckerman, 2003) suggested that the best way to ensure the transfer is to design the classroom learning process as conceptual rather than empirical. Children in a "Vygotskian" classroom start with the core element of a certain subject (e.g., number), and then they create a model of this subject and systematically explore the areas of applicability of this model. Modeling provides the initial generalization so all specific manifestations appear as concrete applications of the core rule. In this way, instead of following an inductive path from concrete manifestations to a (possible) generalization and transfer, students start with a general model that ensures that all possible transfers are already "included" in it.

The last one of the universal criteria of MLE is mediation of meaning. According to Feuerstein (1990, p. 98):

> The mediation of meaning provides the energetic, dynamic source of power that will ensure that the mediational interaction will be experienced by the mediatee. On a more general level, the mediation of meaning becomes the generator of the emotional, motivational, attitudinal, and value-oriented behavior of the individual.

In other words, if the criteria of intentionality and transcendence respond to the question of *how* to create mediated learning interactions, the criterion of meaning responds to the question of *why* we engage in these interactions. When teachers respond "because this is a part of the curriculum" to the students' question "why should we learn this material?" this might be factually true but such an answer is devoid of mediation of meaning and reduces the students' chance of gaining MLE. To experience mediated learning students should understand the motivation behind every step of the educational process.

Mediation of meaning can be performed on different levels but on each one of them teachers should make students aware that learning activities, tasks, and operations are not arbitrary – that they are not a whim of the teacher but represent the necessary steps for turning a student into an independent and self-directed learner. Thus to the question "Why should we learn this material?" one may wish to respond by pointing out how it helps students to develop their abilities, thus leading them to greater independence as self-directed learners and thinkers. On more concrete levels mediation of meaning provides the reason why an element (e.g., an arithmetic operation) occupies a certain position within the whole (e.g., the corpus of mathematical knowledge). Students should be made aware that the specific positioning of this element is not an arbitrary decision of the teachers or textbook authors but reflects certain logic of the given field of knowledge. The teachers' motivation stems from their allegiance with this field as it has developed historically and socioculturally. The mediation of the meaning of teachers' actions does not imply that they should be accepted uncritically. On the contrary, by revealing the motivation behind their actions and on a more general plane the motivation behind the structure of the given field of knowledge, the teachers prompt students to become more critical and reflective and not only of others but also of themselves. The next time the students are asked about their opinion they would probably remember that it is not enough to state that "I just think so," but that one should reflect on the reasons motivating the answer, opinion, or action.

In addition to the three universal criteria of mediation Feuerstein (1990) suggested a number of additional criteria that are contextual and reflect specific needs of children and goals of their mentors. One such criterion is "mediation of the feeling of competence." Students with learning difficulties often suffer not so much from an objective lack of competence as from the *feeling* that they are incompetent. The goal of the mentor is to focus on the positive aspects of the child's performance, emphasize these aspects, and provide elaboration on why the child performed certain operations correctly. Children themselves might be unaware of the reasons for their correct performance. The task of the mentor is to reveal the true meaning of the correct action to the student and use it as a pivotal point for improving other actions that might be problematic. Mediation of the feeling of competence thus contributes to students' motivation to go further – "If I was able to solve this task, for sure I will be able to tackle the next one."

The RMT approach rests on the foundations of MLE theory by infusing all teacher/student interactions with intentionality, transcendence, and meaning. There are two major targets of these interactions: the first is the establishment in the students of efficient cognitive functions of a general nature that are required by any type of systematic learning activity, whereas the second is the appropriation by students of mathematically specific psychological tools.

Development of Cognitive Functions

One of the major claims of the RMT approach is that students' difficulties with mathematical tasks often stem not from the lack of specific mathematical knowledge but from the absence of cognitive prerequisites of a more general nature. To explore this issue one needs a schema for describing cognitive functions that form the basis for efficient problem-solving activity. Contemporary research offers a number of such schemas, most of them based on the notion of information processing (see Anderson, 1996).

Feuerstein, Rand, and Hoffman (1979) and Feuerstein, Rand, Falik, and Feuerstein (2002) used the division of the mental act popular in information-processing approaches into input/elaboration/output phases. Using a variety of clinical observations Feuerstein compiled a list of "deficient cognitive functions" that are often responsible for faulty or inefficient problem solving. In what follows we illustrate how some of the deficient cognitive functions identified by Feuerstein may negatively affect mathematical problem solving.

For example, the lack of spontaneous exploratory activity at the input phase may lead to the lack of encoding of important information given in the mathematical task. In this case the erroneous solution of the problem will be caused not by the lack of mathematical knowledge or operation on the part of the student but by faulty input of data. At the elaboration stage it is often necessary to advance and check several hypotheses. Students who are not used to hypothetical reasoning may resort to simply returning to an inappropriate previous strategy, leading to the wrong solution. In this case, again, the wrong solution comes not from that lack of mathematical knowledge or operation but from a more general problem with hypothetical reasoning. Finally, at the output stage an essentially correct solution found during the input and elaboration phases might be presented by students in such an "egocentric" way that it would be recognized as correct neither by the teacher nor by their peers.

Another cognitive functions schema relevant to our goals was proposed by Sternberg (1980). Rather than arranging functions along the processing axis of input/elaboration/output Sternberg organized them topically and hierarchically into metacomponents: performance components and acquisition components. Metacomponents include identifying the problem, selecting the required lower level operations, choosing the strategy for combining these operations, selecting the optimal representations (e.g., graphic, verbal, and formulaic), and monitoring the process of problem solving. The metacomponents are also responsible for providing the channel of "communication" between performance and acquisition components. Once the problem is identified and plans for solution are selected at the metalevel they should be implemented using the performance components. The latter organize themselves into four major stages of problem solving: encoding the necessary information, executing the working strategy, comparing the obtained solution with available possibilities (e.g., with multiple choice answers), and selecting the response. Acquisition components are specifically related to learning new information and transfer of retained information from one context to another.

For our purpose it is important to emphasize that mathematical problem solving can be impaired by faulty general cognitive functioning in each one of the components but particularly in the metacomponents. Let us illustrate this by the "monk" problem mentioned by Sternberg (1980):

A monk climbed a mountain. He started at 6 A.M. and reached the summit in the evening. He spent the night on the summit. The next morning he arised early and left the summit at 6 A.M. descending by the same route he used the day before and

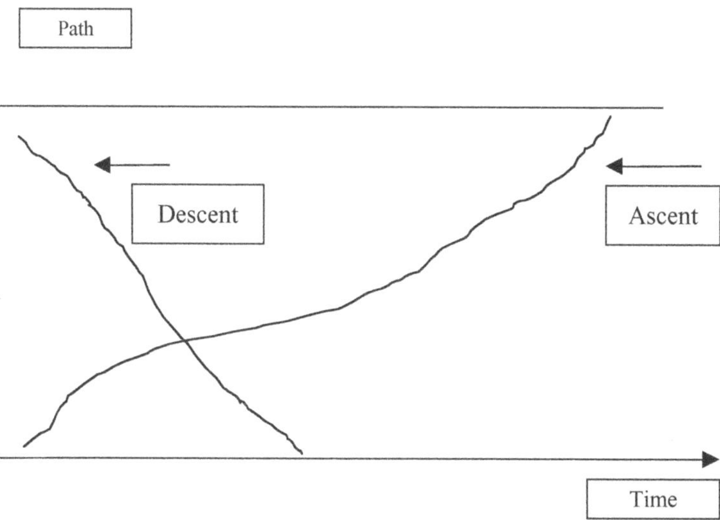

Figure 4.1. The monk problem.

reached the bottom at noon. Prove that there is a time between 6 A.M. and noon at which the monk was at exactly the same spot on the mountain on both days.

The problem is quite difficult if approached "as is," without redefining the problem or without presenting it in a different modality. Efficient reasoning at the metacomponental level, however, would greatly simplify the solution. One approach is to redefine the problem as that of two monks. One of them started to climb the mountain at 6 A.M. while the second one started descending from the top of the mountain at 6 A.M. It now becomes clear that as long as they use the same path there will always be a point where they meet. Another way is to change the representation of the problem from verbal-logical to graphic. Figure 4.1 shows the relevant representation.

The correct solution thus depends not so much on specific physical knowledge or knowledge of mathematical operations but on the general cognitive skills associated with such metacomponents as formulation of a problem and choice of an optimal representation. The latter aspect is of particular importance for the RMT approach because, as we demonstrate later in this chapter, the choice of optimal representation often depends on those symbolic tools that are internalized by students as their inner psychological tools. The problems with the appropriateness of these tools or their internalization may severely handicap the students' ability to create the optimal representation of a mathematical problem.

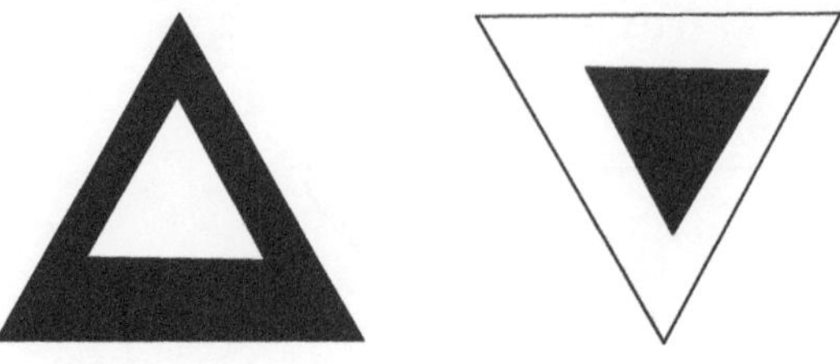

Parameters of comparison:
Shape, Orientation, Color, Color location, Color proportion.

Figure 4.2. Conceptual comparison.

Cognitive Functions in RMT Paradigm

As mentioned in the previous sections, we claim that students' difficulties with mathematical tasks often stem not from the lack of specific mathematical knowledge but from the absence of more general cognitive prerequisites. Below we describe those clusters, networks, or systems of cognitive functions that are essential for constructing mathematical conceptual understanding. In the RMT paradigm we distinguish three broad aspects of a cognitive function – the conceptual component, the action component, and the motivational component – which work in a relationship to each other to provide the cognitive function with its integrity as a distinct mental activity or psychological process. The conceptual component provides a "steering" mechanism to the mental activity by defining or giving description to the nature of the action that is taking place when the function is executed. This component can be further viewed as an interaction between procedure and purpose – the mechanism of the function as it is guided by and shaped through a conceptual meaning. One example relates to the cognitive function comparing, whose conceptual meaning is the idea of similarities and differences between or among two or more objects or events (see Figure 4.2). The operational mechanism for *comparing* is to first select and define a quality, dimension, or concept by which to compare and then characterize separately each object or event to be compared based on this quality or dimension and then, finally, look for the similarities and differences between or among the characterizations of the objects. Thus each one of the two objects in Figure 4.2 should first be analyzed in terms of the chosen parameters and only then compared. Thus, operationally, this process of *comparing* is the integration of a mental procedure with a distinct conceptual understanding.

In addition to conceptual and operational components there is also a motivational aspect of cognitive function. The motivational component stems from the learner's awareness of the perceived importance or benefits of carrying out this mental action. The motivational component is developed as the learner is guided to practice the use of the conceptual know-how of the cognitive function through various modalities and at various levels of complexity and abstraction while being clearly and explicitly made aware of the benefits of this action through the mediation of its meaning and a feeling of competence. The motivational component of cognitive functions thus emerges as a function-bound energizing quality that becomes an integral aspect of the function.

There are three levels of cognitive functions required for rigorous mathematical thought (see Table 4.1). The first level consists of general cognitive functions needed for qualitative thinking in dealing with any content or task. Before most learners are engaged in rigorous conceptual reasoning their cognitive processing occurs mainly at the concrete level and is dominated by already existent natural psychological functions (see Vygotsky, 1998). The learners' interaction with the world around them focuses mostly on empirically given familiar objects and events. The cognitive functioning that emerges is shaped by the learners' everyday spontaneous concepts that can be experientially rich but are usually episodic, unsystematic, and nonrigorous. Systematic mediation of the learners' cognitive functions facilitates the development of what Vygotsky called "scientific" concepts that are systemically organized and rigorous. In RMT theory this transformation is facilitated through connecting the learner's everyday spontaneous concepts to the operational conceptual know-how of the Level 1 cognitive functions. By leading learners through a series of cognitive tasks focusing on comparison the mentor connects the preexistent notions of "same" and "not same" or "different" to the conceptual cognitive function of *comparison*. The conceptual approach to comparison elevates it from the level of empirically perceptible "A is bigger than B" to such concepts as size, for example, "A and B are different in size."

The second level consists of cognitive functions that are required for quantitative thinking and precision. These functions have more structure than the general cognitive functions because they share a greater interrelatedness through the conceptual basis of quantity, whereas the general cognitive functions serve as a necessary platform for the construction of quantitative thought. The third level of cognitive functions integrates processing regarding quantity and precision into a unique fabric of logic and generalized abstract relational thinking needed specifically for the mathematics culture. Together, these three levels of cognitive functions define a range of mental processing

Table 4.1. Three levels of cognitive functions for RMT

Cognitive function	Definition
Level 1 – General cognitive functions for qualitative thinking	
1. Labeling-visualizing	1. Giving something a name based on its critical attributes while forming a picture of it in the mind or producing an internalized construction of an object when its name is presented.
2. Comparing	2. Looking for similarities and differences between two or more objects, occurrences, or situations.
3. Searching systematically to gather clear and complete information	3. Looking in a purposeful, organized, and planningful way to collect clear and complete information.
4. Using more than one source of information	4. Mentally working with two or more concepts at one time, such as color, size, and shape, or examining a situation from more than one point of view.
5. Encoding-decoding	5. Putting meaning into a code (symbol or sign) and/or taking meaning out of a code.
Level 2 – Cognitive functions for quantitative thinking with precision	
1. Conserving constancy	1. Identifying and describing what stays the same in terms of an attribute, concept, or relationship while some other things are changing.
2. Quantifying space and spatial relationships	2. Using an internal and/or an external system of reference as a guide or an integrated guide to organize, analyze, help articulate, and quantify differentiated, representational space and spatial relationships based on whole-to-parts relationships.
3. Quantifying time and temporal relationships	3. Establishing referents to categorize, quantify, and order time and temporal relationships based on whole-to-parts relationships.
4. Analyzing-integrating	4. Breaking a whole or a decomposing a quantity into its critical attributes or its composing quantities – constructing a whole by merging its parts or critical attributes or composing a quantity by merging other quantities together.
5. Generalizing	5. Observing and describing the nature or the behavior of an object or a group of objects without referring to specific details or critical attributes.
6. Being precise	6. Striving to be focused and exact.

Cognitive function	Definition
Level 3 – Cognitive functions for generalized, logical abstract relational thinking in the mathematics culture	
1. Activating prior mathematically related knowledge	1. Mobilizing previously acquired mathematical knowledge by searching through past experiences to make associations and coordinate aspects of something currently being considered and aspects of those past experiences.
2. Providing and articulating mathematical logical evidence	2. Giving supporting details, clues, and proof that make mathematical sense to substantiate the validity of a statement, hypothesis, or conjecture. Generating conjectures, questions, seeking answers, and communicating explanations while complying with the rule of mathematics and ensuring logical consistency.
3. Defining the problem	3. Looking beneath the surface by analyzing and seeing relationships to figure out precisely what has to be done mathematically.
4. Inferential-hypothetical thinking	4. Forming a mathematical proposition or hypothesis and searching for mathematical logical evidence to support the proposition or hypothesis or deny it. Developing valid generalizations and proofs based on a number of mathematical events.
5. Projecting and restructuring relationships	5. Forming connections between seemingly isolated objects or events and reconstructing existing connections between objects or events to solve new problems.
6. Forming proportional quantitative relationships	6. Establishing a quantitative relationship of correspondence between a concept (or a dimension) A and a different concept (or a dimension) B or between the same concept in two different contexts by (1) determining some original amount of A and a connecting original amount of B and (2) hypothetically testing to see that for any multiples of the original quantity A the corresponding quantities of B will result from the same multiples of the original quantity of B.
7. Forming a functional relationship	7. Making connections between two or more things that are changing their values in such a way that the changes form a network or work together in an interdependent way.

(continued)

Table 4.1. (*continued*)

Cognitive function	Definition
8. Forming a unit functional relationship	8. Making a connection between the change in the amount of the dependent variable that is produced by a unit change in the amount for the independent variable that is defined by the functional relationship between the two variables expressed in the mathematical function or the algebraic equation.
9. Mathematical inductive-deductive thinking	9. Taking aspects from various mathematical details that seem to form a pattern, categorizing them into general relationships of attributes and/or behaviors, and organizing the results to form a general mathematical rule, principle, formula, recipe, or guide; applying a general rule or formula to a specific situation or a set of details that connect only with the rule in terms of belonging to categories of attributes and/or behaviors expressed by the rule.
10. Mathematical analogical thinking	10. Analyzing the structure of both a well-understood and a new mathematical operation, principle, or problem, forming relational aspects of the components of each structure separately, mapping the set of relationships from the well-understood structure to the set of relationships for the new structure, and using one's knowledge about the well-understood situation along with the mapping to construct understanding and insight about the new situation.
11. Mathematical syllogistic thinking	11. Using the relationship established between item A and item B stated in a mathematical proposition along with the relationship established between item A and item C stated in a second mathematical proposition to logically infer a previously unknown relationship between item B and item C.
12. Mathematical transitive relational thinking	12. Considering a mathematical proposition that presents a quantitatively ordered relationship ($>$, $<$, $=$, etc.) between two mathematical objects A and B along with a second mathematical proposition that presents a quantitatively ordered relationship between mathematical objects A and C and then engaging in inferential deductive thinking to logically transfer a quantitatively ordered relationship between objects B and C.
13. Elaborating mathematical activity through cognitive categories	13. Reflecting on and analyzing mathematical activity and discovering, labeling, and articulating, orally and in writing, underlying mathematical principles and concepts using the language of mathematics and cognitive functions

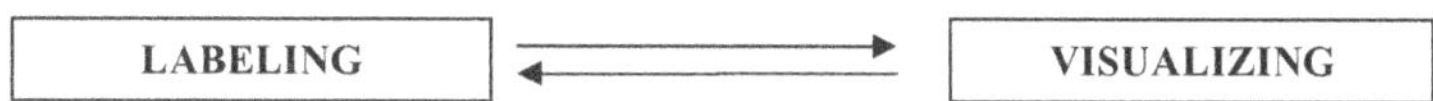

Figure 4.3. Schematic of the cognitive function labeling-visualizing.

that extends from general cognitive skills to higher order mathematically specific functions.

In Table 4.1 a number of cognitive functions have hyphenated names and may appear to be two functions combined together. There are two reasons we have taken this approach. First, we contend that in such cases, as for *labeling-visualizing, encoding-decoding, analyzing-integrating,* and *mathematical inductive-deductive thinking,* the operational component of the function exists in a spectrum of conceptual opposites that equip the learner with a more robust structure of conceptual know-how (see Figures 4.3 to 4.6). For example, when the cognitive function *labeling-visualizing* (see Figure 4.3) emerges in the learners they are giving an object a name or forming a picture of the object in their mind with both linked through identification of the critical attributes of the object. The second reason is that when a hyphenated cognitive function becomes fully crystallized in the learners' minds the conceptual know-how of this function is amplified by the copresence of conceptual opposites. When labeling-visualizing is fully crystallized in the learners and when they seek to give an object a name, they spontaneously form a picture of the object in their mind and vice versa. Such duality of hyphenated function serves to strengthen the systemic character of cognitive functions.

As analyzing and integrating are not separate cognitive actions, adding and subtracting are not separate mathematical events. They are unified complementary actions and events. Adding and subtracting are linear mathematical actions.

Development of cognitive functions on all three levels creates a pathway between general cognitive functions that tap into the learners' everyday spontaneous concepts and the "scientific" concepts of mathematics. This is a two-way process that was described by Vygotsky (1986) in terms of learners' zone of proximal development (ZPD). "Scientific concepts move from the 'top' downward – from verbal-logical formulae to concrete material. Spontaneous concepts move in the opposite direction, from the 'bottom' upward – from contextual everyday experience to the formal structures of well-organized thought" (Kozulin, 1998a, p. 49). We propose that the process of mediating

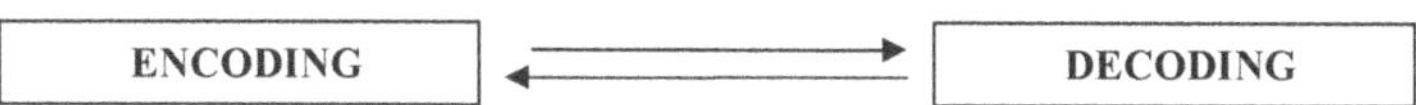

Figure 4.4. Schematic of the cognitive function encoding-decoding.

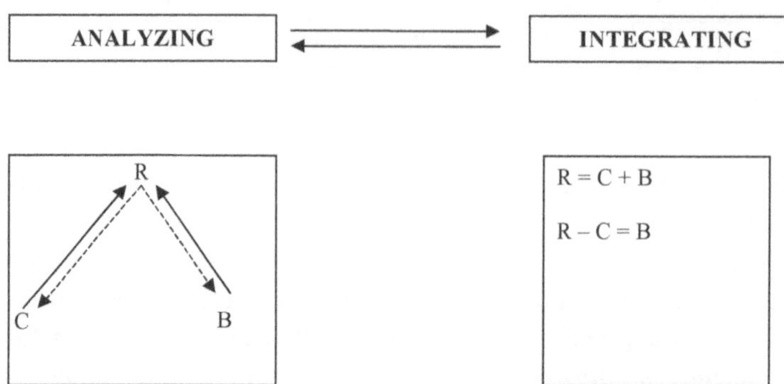

Figure 4.5. Schematic of the cognitive function analyzing-integrating.

the development of the three levels of cognitive functions in the learner creates a broad zone of proximal development, which we view as a pertinent aspect for effective mathematical learning in the classroom. In Chapter 6, we examine the difference between students' actual level of mathematical understanding and their potential level of mathematical structural development. Within the ZPD the three levels of cognitive functions are in their formative states and are emerging.

Instrumental Enrichment Program

One of the main applications of Feuerstein's theory of mediated learning experience is a cognitive enrichment program called Instrumental Enrichment (IE) (Feuerstein, Rand, Hoffman, and Miller, 1980). The IE program was initially developed for application with culturally different and socially disadvantaged adolescents in Israel, but since the mid-1980s it has been translated into a number of European and Asian languages and used with various populations of learners ranging from children with disabilities to high-functioning young adults (see Kozulin, 2000).

The main objective of the IE program, as stated by Feuerstein, is to increase human modifiability, making learners more amenable to direct learning situations. Thus although the application of the IE program is saturated with mediated learning, the goal of this program is to create in the learner the preconditions for efficient direct learning. The subgoals of IE include remediation

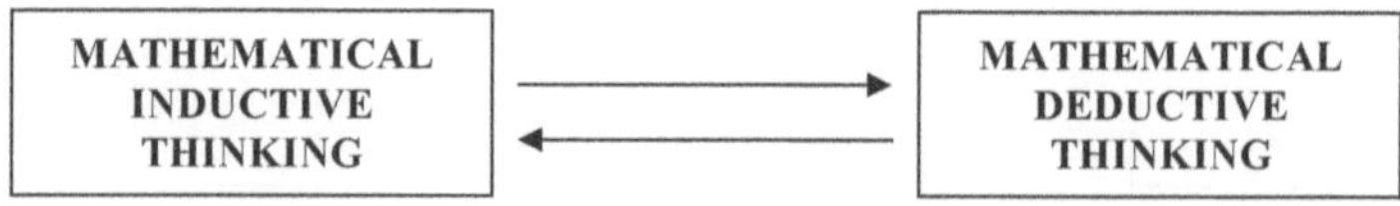

Figure 4.6. Schematic of the cognitive function mathematical inductive thinking-mathematical deductive thinking.

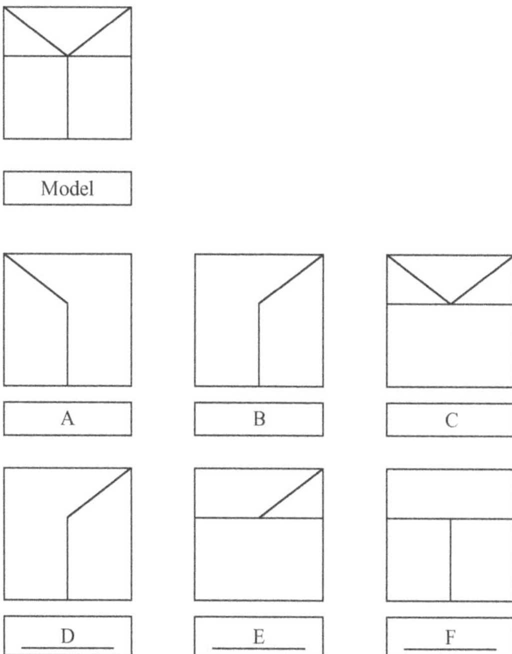

Figure 4.7. Find two frames that together contain all elements of the model. Respond by circling letters under relevant frames.

of deficient cognitive functions, enrichment of learners' concepts and operations, improvement of intrinsic as well as task-related motivation, and enhancement of reflective thinking and metacognitive skills. Thus the IE program as a whole aims at helping to turn learners from passive recipients of information into active constructors of the new knowledge (see Feuerstein et al., 1980, pp. 115–118).

One may distinguish two major aspects of the IE program: the tasks themselves and the didactics of mediating them to the learners. The tasks of the IE program are organized into 14 booklets or "instruments" that cover such cognitive areas as analytic perception, orientation in space and time, comparison, categorization, syllogistic reasoning, and so on. The students' classroom learning of the IE program is mediated by teachers specially trained in the philosophy and technique of this program.

The task presented in Figure 4.7 can be used for illustrating some of the features typical for the IE tasks.[1]

[1] In order to preserve the integrity of the IE program material we used in this chapter the illustrative tasks conceptually similar but graphically different from those of the IE program itself.

First, like any proper cognitively oriented task IE tasks are constructed with the aim at the process rather than product. The comparison of geometric elements thus constitutes means rather than the goal of activity. The true goal is to engage the student in the cognitive process that includes the following steps. (1) Taking stock of the available data and labeling these data. [In the present task these data appear as a model, frames with geometric elements that can be used for solving the problem, and verbal instruction.] (2) Formulating the problem. [To formulate the problem students often need to perform integration of different sources of information. In the present case this requires integration of graphic information of the models and response frames with verbal information of the instruction.] (3) Developing the problem-solving plan and choosing the starting point. [In the present task students should start with a thorough analysis of the model in terms of its constituent elements. The problem-solving plan may include such specific strategy as eliminating pairs of frames that contain identical rather than complementary elements. For example, the pair of frames A and B cannot be considered as a candidate for a correct solution, not only because they together do not contain all the necessary elements but also because they have a common element – the vertical segment.] (4) Implementing the problem solving plan and arriving at the solution. [The analysis of the model leads to identification of the four elements: vertical segment, horizontal segment, and two diagonal segments. The plan may include either a systematic comparison of frames for the presence of complementary elements or the preliminary elimination of pairs that contain a major common element, for example, a vertical segment (A, B, D, and F). Such elimination may narrow the search to just few combinations: A-E, B-E, C-E, and F-E.] (5) Once the correct answer A-E is found, it should be checked against the model and only then written down.

The role of the teacher is to lead students toward greater awareness of their problem-solving actions while supporting them in the problematic points. For example, certain difficulties may emerge as early as at the stage of defining the problem. Some students may claim that the pair of frames C and F provides the correct answer because together they contain all elements of the model. Here the role of the teacher is to explore, together with students, the explicit and implicit meaning of the instruction that asks to find "two frames that together contain all elements of the model." The expression "contains all elements" explicitly states that all elements of the model should be present, but it also implicitly stipulates that there should be no additional elements. Thus the pair C and F is unsuitable because it contains two horizontal segments. Once the principle of implicit instruction is established and accepted by the students, the teacher may start asking students about other instances when they have confronted implicit instructions. Such an inquiry serves an important role in "bridging" cognitive principles learned with the help of IE tasks to the curricular material and everyday life situations. If students experience difficulty in coming up with the examples of implicit instruction, the teacher may confront them with the following mathematical task and ask them to look for implicit instruction:

Find the values of X that satisfy the following algebraic equation: $X^2 = 9$.

Students are expected to come up with the explanation that the implicit instruction in this task is to identify all suitable values of X (i.e., both 3 and -3). It would be particularly rewarding if some of the students point out that the word "values" (in plural) provides a cue to the nature of implicit instruction. The notion of "cues" is one of the cognitive terms that students are expected to acquire during the IE lessons and start using in the analysis of their own problem solving.

Thus, for example, at the stage of planning the students are encouraged to state their selection of a strategy. If one of the students suggests counting elements in each of the frames (Figure 4.7), he or she should be asked in which sense this is a strategy. If the student responds that the model consists of four elements and thus the two frames are expected to have $2 + 2$ or $1 + 3$ elements, such an answer should be encouraged because counting elements is indeed a strategy. After this initial encouragement, however, students' attention should be drawn to the fact that five of six frames have two elements each and thus, though counting elements is a strategy, it cannot be accepted as an efficient strategy.

Another illustration of a problem similar to IE tasks is presented in Figure 4.8. At first glance this task has very little in common with that shown in Figure 4.7. One of them requires analytic perception of geometric shapes, whereas the other is based on systematic comparison in different modalities (pictorial, graphic, and verbal). And yet, the work associated with these two tasks is quite similar. Such a connection between different tasks is an important "built-in" feature of the IE program. The program is organized as a system: the same requirements for cognitive functions, operations, strategies, and cognitive principles appear again and again in different "instruments" of this program. To discover this interconnectedness one should go beyond the appearance of the tasks and delve into their cognitive aspects. For example, in both tasks (Figures 4.7 and 4.8) analyzing the model plays a very important role. As in Figure 4.7, one cannot plan the solution before the model is analyzed into its constituent geometric elements; in a similar way, a new object in Figure 4.8 cannot be created unless the initial object is properly analyzed and described as, for example, "one female figure approximately 15 mm in size."

Object	Parameters of change	New object
	color	
	gender, number and size	
"red house"	adjective	

Figure 4.8. Create new objects using given parameters of change. Respond by drawing or writing your answers.

Another common aspect of the tasks is that in both of them the learner is confronted with both explicit and implicit instructions. We have already discussed that in the first task (Figure 4.7) it is explicitly stated that two of the chosen frames should have all the elements of the model, but it is also implicitly "stated" that there should be no additional elements. In the second task (Figure 4.8) there is an explicit instruction to create a new object by transforming the original one using given parameters of change, but in addition there is also an implicit instruction to change no other parameters. So, for example, by changing the color of the black square one is expected to preserve its shape, size, number, and orientation.

Finally, there are always some basic operations that reappear in different IE tasks. Thus the last stage of the problem-solving process – the choice of the answer and verification of its correctness – in both cases requires comparison of models. The difference is that in the first task (Figure 4.7) the elements of the chosen frames should be compared to the elements of the model, whereas in the second task (Figure 4.8) the unchanged parameters of the new object should be compared to the original object while the changed parameters should be compared to the information provided in the middle column. As in all other IE tasks the teacher should ultimately lead the discussion about the required operations beyond the IE tasks and into the curricular field. Thus the students might be asked to identify the operation of comparison in the final stages of mathematical problem solving when the answer is compared to that initially given.

The major criteria of MLE (intentionality, transcendence, and meaning) are always kept in mind by the IE teacher when applying the program. The amount of support provided to students, the size of a problem-solving step, and even the number of tasks solved during the lessons all reflect the teacher's intention to make the program suitable to the needs of a given group of students. The considerable freedom given to students in collectively defining, interpreting, and suggesting solutions to the tasks mediates the meaning of the IE problem-solving activity, which is coconstructed by students and their teacher. Finally, "bridging" from the cognitive aspects of the tasks to the cognitive aspects of other curricular subjects reflects the transcendent nature of MLE.

In what follows, however, we focus on one aspect of the IE program that was not explicitly stated by its authors (Feuerstein et al., 1980) and yet, in our opinion, constitutes one of its stronger features. We are talking here about the IE program as a potentially rich system of psychological tools.

IE Program as a System of Psychological Tools

The notion of psychological tools (see Chapter 3) is one of the central concepts of Vygotsky's sociocultural theory. Each culture develops its own set of symbolic tools that are appropriated by its members. To a considerable extent the transmission of culture from generation to generation depends on the transmission of these systems of tools. Signs, symbols, writing, formulae, graphs, maps, and pictures are just some of the better known symbolic tools. There are two stages in the transformation of external symbolic tools into inner psychological tools. First, the tool should be identified within the content material and appropriated. For example, a plan of the city should be first appropriated as an external symbolic tool helping to orient oneself in the given geographic area. After that the symbolic aspects of the plan should be internalized as an inner psychological tool that allows the person to comprehend his or her environment beyond what is given by his or her senses. Psychological tools help people to "see" beyond what is given; they turn their personal experience into more generalized schemata that create conditions for sharing these experiences with others. For example, people who do not have a map of a city are confined to the streets and buildings they see at any given moment. Not knowing the scale of, say, New York City, they cannot relate their experience ("I walked four blocks") to the goal, for example, to reach Wall Street. They also cannot convey information to others because they lack a common schematic, including reference points. Efficient orientation in space, however, does not rely only on the use of actual plans and maps but on the inner psychological tools that lead us beyond our immediate experience into a schematic, generalizable, and transmittable model of the environment.

Vygotsky (1979) emphasized that appropriation and internalization of psychological tools constitute an essential goal of education. So-called content of learning reveals its true meaning only when students approach it with the help of relevant symbolic and psychological tools. The acquisition of symbolic artifacts as tools may constitute, however, a considerable problem, especially for underprivileged and minority students whose immediate environment does not support such an acquisition in a "natural" way. Let us consider two hypothetical families. In one of them the *Wall Street Journal* not only constitutes the parents' everyday reading material, but the parents also take the time to explain to their children the meaning of some of the graphs that appear in the paper. In the second family, a newspaper is a relatively rare object, but even when the father brings it in he would never consider sharing with his children the meaning of the tables that appear on the sports pages. One may easily guess which children would experience greater difficulty in mastering materials of the modern textbooks that are crammed with tables, graphs, and diagrams.

However, even students from more privileged families often experience difficulties with acquisition of symbolic tools. These difficulties are inherent in the educational systems that focus on content, such as historical facts, literary texts, or mathematical rules, without paying attention to those tools that help students organize these facts, texts, and rules in their heads. These difficulties also come from teachers' lack of skills in presenting symbolic artifacts that appear in curricular material as tools rather than a part of content. In this respect the IE program is a rare exception, because it consistently and systematically provides the necessary basis for acquisition and internalization of symbolic tools required in practically all school subjects.

First, we will go over some of the symbolic tools that can be acquired via the IE program, such as codes, tables, and diagrams. Though coding (i.e., substitution of a simple sign, for example, a letter or a digit) for an object or a concept seems to be a relatively simple symbolic operation, it is absolutely central for efficient problem solving in mathematics and science. Of course, coding in the broader sense of symbolic substitution is ubiquitous in human life. Any spoken or written word is based on coding, but we are rarely aware of this because we perceive our native language and script as "natural." There is, of course, nothing natural here. The string of signs h-o-u-s-e has no "natural" connection to "house" as an object or a concept. And yet, because of this illusion of naturalness, students usually start to experience difficulties when coding goes beyond their native language and into the sphere of artificial codes that designate objects, concepts, or statements. The IE program systematically introduces different codes with a goal of teaching the students how to distance themselves from concrete objects and start operating with their more abstractive representations. One of the simpler forms of coding is labeling. For example, in a given task that includes six objects (e.g., cylinders) of three different colors and two different sizes, we can label them A, B, C, a, b, and c, where letters indicate different colors, whereas the letters' sizes (regular or capital) indicate object's size (see Figure 4.9). Instead of returning each time to actual physical images of these cylinders students may start engaging in different operations of comparison and classification using codes rather than objects.

One of the pages in the "Orientation in Space" booklet of IE is devoted exclusively to the exploration of different forms of symbolization and coding, from representations maximally close to the object, such as footprints, to those that are more remote but still somewhat perceptually connected to the objects, such as international road signs (e.g., "no left turn"), and then to fully conventional signs such as mathematical operators. Students thus learn that certain types of coding (e.g., using arrows to indicate the direction) require symbols that have certain properties, such as asymmetrical directionality of an arrow, whereas other symbols are purely conventional, such as those for multiplication and division. Other exercises in IE teach students to use coding in a variety of contexts, such as to indicate the type of mistake, position of an object, set of objects to be classified, or even a place for answer that is conventionally represented by an empty line______.

The diversity of contexts leads students to the development of a general notion of coding rather than just remembering that "x" usually signifies an argument, whereas "y" usually signifies a function. Try asking your students whether we can use "a" to signify an argument and "b" to signify a function, and you would see how reluctant they are to admit that the use of specific codes is arbitrary and based only on mutually accepted convention.

Another important symbolic tool widely represented in the IE program is a table. In Chapter 3 we showed that even in more advantaged students the acquisition of a table as a symbolic tool rarely happens spontaneously. Moreover, even teachers demonstrated difficulty in using a table as an inner psychological tool. At the same time, one can hardly overestimate the role of a table as a cognitive tool connecting data input and data elaboration.

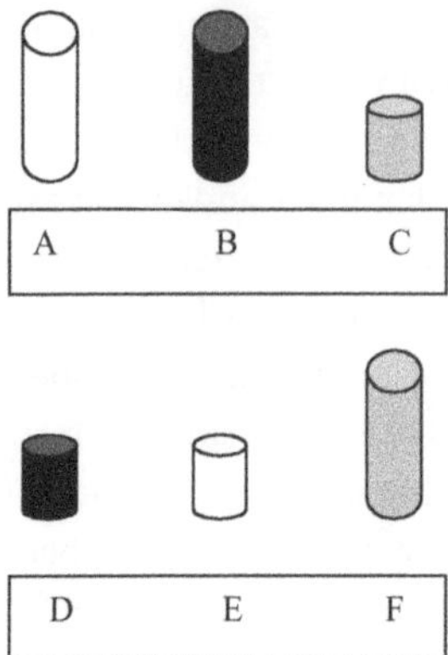

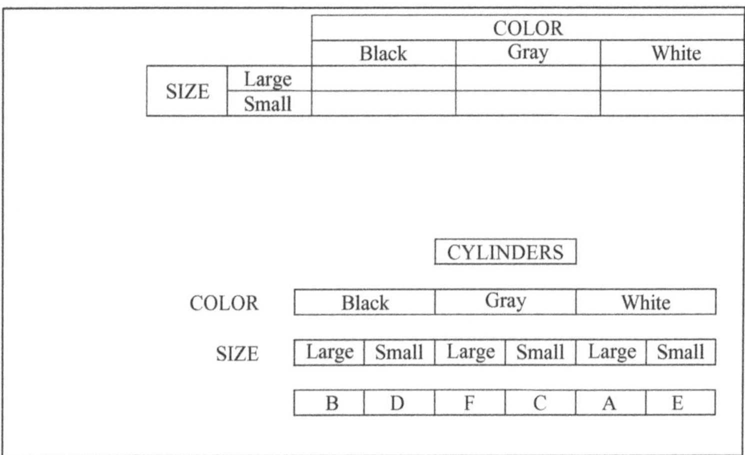

		COLOR		
		Black	Gray	White
SIZE	Large			
	Small			

CYLINDERS

COLOR	Black		Gray		White	
SIZE	Large	Small	Large	Small	Large	Small
	B	D	F	C	A	E

Figure 4.9. Classification task.

The relationship between the structure and function of a symbolic tool is particularly clear in the case of a table. There are two major functions of a table: the first is to organize data according to superordinate categories and the second is to separate knowns from the unknowns in problem solving. Table headings support the first function, whereas the combination of filled and empty cells supports the second. Let us look again at the task in Figure 4.8. Students who have mastered the table as a tool would have no problem answering the following question: "If you continue this table what would you add to the left column?" They would say that it could be practically any picture, graphic image, or text. As for the second column, the answer

Object	Parameters of change	New object
■		□
	adjective	"white house"

Fill in what is missing.

Figure 4.10. Reconstructing original objects.

would be "any parameter of change relevant to the object in the first column." The tabular format focuses students' attention on the issue of relevance by providing an easily observed connection between a superordinate concept of "parameter of change" and specific limitations imposed by the given object. Not every object can be changed along any parameter of change, for example, if your object is a circle you cannot change its "direction." By providing a tool for keeping the parameter and the object side by side a table facilitates the identification of relevant parameters of change that otherwise may become rather complicated.

Finally, the empty cells in the right column of Figure 4.8 help to separate the knowns from the unknowns in problem solving. This aspect is systematically developed in the IE tasks that use what can be called "irregular tables." Figure 4.10 demonstrates such an irregular table. In the first task the parameter of change should be found, whereas in the second task the original object should be reconstructed on the basis of the given new object and the parameter of change. Systematic work with such irregular tables in different IE instruments provides students with the cognitive tools to analyze any problem into given data and required actions. For example, in the third task in Figure 4.10 only the new object is given. It is impossible to reconstruct the original object without adding some parameters of change. If one adds "number" as the parameter of change, then the original object would be two or more trees, but if the added parameter of change is "size," then the original object would be one tree but either smaller or bigger than that given as the new object.

During IE lessons students are taught to use a table as a tool for identifying the type of problem. For example, if in the table the first column provides information about some object and the second one provides information about the parameter of change, then the students may guess that the number of possible answers (possible new objects) will be smaller than in the case where only the parameters of the original object are given and both the parameters of change and those of the new object should be determined.

Teaching how to think with superordinate concepts is one of the major goals of a table as a psychological tool. In Figure 4.8 only the columns have superordinate headings, but other tables have superordinate headings both in columns and in rows. As a result the data located in a given cell can always be analyzed and classified according to two categories. One may test the students' degree of internalization of a table as a psychological tool by asking them to suggest an optimal format for presentation of certain data. Thus the task of classifying the cylinders shown in Figure 4.9 requires involvement of two superordinate categories – color and size. The optimal representation then would be a 3×2 table with two major categories/headings (color and size), one of them determining the content of columns and the other one that of rows. The color heading then should be subdivided into black, gray, and white, whereas the size should be subdivided into small and large. This mental schema of two superordinate categories and a six-cell table should appear in the students' minds as appropriate for the optimal representation of given task classification.

When a table fails to be internalized as an inner psychological tool students experience difficulty in choosing the proper representational format. Kozulin (2005b) demonstrated that not only students but also a considerable number of teachers chose a nonoptimal tabular format for representation of a classification task of a group of numbers into one-, two-, and three-digit, as well as odd and even, numbers. Such a task calls for essentially the same 3×2 table format as discussed with regard to Figure 4.9, because it involves two superordinate categories, odd/even (two columns or rows) and number of digits (three columns or rows). Some teachers, however, produced tables in which both categories were presented side by side in the column headings (i.e., odd, even, one-digit, two-digit, and three-digit), producing a five-column table with no rows. As a result the same information (numbers) appeared twice in two different columns, whereas some numbers were lost. After being trained in the IE program the teachers significantly improved their ability to create task-appropriate tables.

Internalized tabular thinking supports such an important cognitive function as selection of an appropriate number of parameters for the analysis of a situation. For example, in the task of comparing the cost of a certain product in three different supermarkets located at different distances from the home of the buyer, students whose thinking does not include psychological tools may focus on only one parameter, the cost of product, and ignore the second one, the cost of travel. The mental tabular schema helps to recognize this as a 3 × 3 task with two superordinate categories (cost of product and cost of travel).

Instrumental Enrichment and Math Problem Solving in Immigrant Students

Immigrant children constitute one of the student groups who have special needs with respect to learning mathematics. This applies to all immigrants irrespective of their origin for at least two reasons: (1) The language component is far from negligible in understanding math tasks. Moreover, the academic language used in math textbooks differs from everyday conversational language. Many immigrant children who are conversationally fluent in their new second language still experience serious difficulties with the comprehension of academic language (Cummins, 2000). (2) There are always differences among instructional approaches typical for various national educational systems. Thus even when the contents of math curricula are similar, immigrant students still face the challenge of adjusting to a new instructional approach.

Using the results of the Third International Mathematics and Science Study (TIMSS) Huang (2000) analyzed math achievement of primary school immigrant children in five English-speaking countries. In the United States, Canada, and England, first-generation immigrants proved to lag considerably behind their nonimmigrant peers. Though this gap narrowed for the second-generation immigrant students, it still remained significant for the U.S. and Canada samples. As could be expected, the greatest disadvantage was observed in the first-generation immigrants whose home language was not English. Two countries in which first-generation immigrants showed no math disadvantage are Australia and New Zealand. Immigration laws in these countries favor professional immigration and thus their immigrants have, on average, a higher educational level. However, the second-generation Australian immigrants whose home language was other than English showed significantly lower results than the first-generation immigrants who spoke English at home. These data based on representative national samples confirm that immigrant students long remain at risk of disadvantage when it comes to learning math. It also indirectly points to the importance of the educational level of the parents. This brings us to the issue of two different types of immigrant children: those who come with kindergarten or school experience broadly compatible with that of the new country and those who either have no previous formal education or whose educational experiences differ enormously from those of the modern classroom. One of us (A.K.) vividly remembers his visit in the mid-1990s to a primary school in one of the South African black townships. About 50 students were sitting absolutely quietly in a crowded and airless classroom. In front of each one of the students was an open notebook whose pages were covered by an almost calligraphic handwriting. After a quick examination it became clear that the texts in all notebooks were similar and corresponded to the text that the teacher wrote on the blackboard. When later he inquired about other classroom activities, his question created considerable puzzlement on the part of the teacher, who reiterated that "students are learning how to write." When he asked directly whether there were any activities other than copying from the blackboard, the answer was no. It is easy to imagine that this type of educational experience based on rote learning may create a considerable problem for children who become immigrants or even migrants to more affluent areas of their own country and find themselves in the modern classroom that emphasizes students' initiative and process-oriented problem solving.

In what follows we describe the results of two studies conducted with immigrant students from Ethiopia in Israel (Kozulin, 2005a; Kozulin, Kaufman, and Lurie, 1997). Though each immigrant group and each receiving educational system have their own characteristic features, we believe that the encounter between Ethiopian immigrants and Israeli education revealed some of the universal problems inherent in the integration of students from traditional, mostly preliterate, societies into Western formal educational system as well as the ways of resolving these problems.

Study 1

Students who participated in this study arrived in Israel from Ethiopia when they were 10–11 years old. For the most part they came from rural areas where formal schooling was rare. As a result the majority of new immigrants were illiterate in their native Amharic language. On arrival the students were placed into intensive Hebrew study groups for a period of 3 to 6 months and after that integrated into regular classes according to their chronological age. For about 3–4 years they were studying in regular classes while receiving additional Hebrew lessons to which all immigrant students in Israel are entitled. At the end of this 3- to 4-year-long period it became clear that though the students demonstrated sufficient progress in oral Hebrew communication they still lagged considerably behind their native-born peers in academic Hebrew literacy and curriculum subjects, including mathematics. Such underachievement jeopardized their chances for successful high school graduation. At this juncture the educational authority responsible for immigrant students decided to offer the students at risk an opportunity to join an intensive residential program that would boost their learning and give them a chance to meet the standards of high school matriculation. The residential option was chosen because many of the immigrant families had a very low socioeconomic status and could not afford to provide their children with educational enrichment activities. In the past residential programs for immigrant students from North Africa proved to be successful in advancing their educational achievement (see Feuerstein et al., 1980, Chapter 10).

Immigrant students were placed into experimental classes, 15–20 students per group. A special curriculum was designed that included 12 hours per week of Hebrew, 6–7 hours of mathematics, and 4–5 hours of the IE program. Other subjects were taught according to the standard program, but all teachers received an introduction into the principles of mediated learning and IE. Below we report the results of the first year of the program implementation. At the beginning of the program all students were tested in mathematics and reading comprehension. In addition, their nonverbal cognitive performance was assessed by use of Raven Standard Progressive Matrices test (see Kozulin, Kaufman, and Lurie, 1997). The results of the pretests confirmed that students lagged behind their native-born peers not only in reading and mathematics but also in nonverbal problem solving. This result confirmed our hypothesis that the students' slow progress in curricular areas was determined, at least in part, by the lack of more general cognitive and learning strategies rather than just inadequate subject knowledge or insufficient language skills. The gap between immigrant students' cognitive performance and the Israeli age norms ranged from one to two standard deviations. At the same time, the students revealed considerable learning potential when assessed with the help of the dynamic form of a cognitive matrices test (Feuerstein, Rand, and Hoffman, 1979) that includes an active mediated learning phase. Thus it was demonstrated that the standard cognitive performance of immigrant students does not predict their learning ability. When provided with mediated learning, the students were able to quickly acquire more efficient problem-solving strategies and implement them with relatively difficult nonverbal tasks. The change was not only quantitative but also qualitative – in the dynamic test students demonstrated a profile of responses much closer to that of native-born peers (Kozulin, 1998a). In view of such pretest results it was decided that the IE

Table 4.2. Pre- and posttest mathematics scores of immigrant students (N = 56); standard deviation in parentheses

Pretest	62.7 (16.9)
Posttest	79.4 (15.7)*

*$t = 9.1$; $p < 0.01$; effect size = 1.01.

program would be positioned at the conceptual center of the program with subsequent "bridging" of general cognitive principles to specific curricular areas.

Table 4.2 shows the results of the 8-month-long intervention as it pertains to mathematics. The material of both the pre- and posttest included such basic mathematical problem-solving tasks as arithmetical operations, simple fractions, and text problems. An effect size of 1.02 is large. (According to Cohen, 1988, an effect size of 0.5 is considered moderate and an effect size of 0.8 is large.) The majority of students who for years were unable to master the basic mathematical problem-solving skills successfully realized their latent learning potential and in one school year reached the level of almost complete mastery of basic mathematics. At the same time it should be taken into account that basic mathematical problem-solving skills are the necessary, but not sufficient, basis for successful high school studies. The above program relied mostly on more general cognitive strategies provided by IE and less on mathematically specific psychological tools. In this we see one of the differences between the use of IE in its "pure" form and its integration within the RMT approach.

Study 2

Immigrant students who participated in this study came to Israel from Ethiopia when they were 4–5 years old. Their formal education started in Israeli kindergartens and primary schools. With such an early start one might expect a rather smooth integration into the educational system, and yet at the age of 10–11 (4th grade) these students still experienced serious problems with reading comprehension and the most basic mathematical skills. Supplementary Hebrew lessons seemed to be insufficient for bringing their achievement to the grade-appropriate level. The experience of failure caused these students to become passive and the gap between them and the class widened. As a result some of them were referred to placement committees and recommended for transfer to special education classes. In response to this need a special program called CoReL (Concentrated Reinforcement

Lessons) was created (Kozulin, 2005a). The principles of CoReL include the following:

- The provisional nature of the program – a preset maximum period of activity both for the student and the teacher (4 months to one year). Once the student reached the benchmark achievement level he or she leaves the CoReL.
- Integration of general cognitive and domain-specific learning skills. CoReL includes both IE and domain-specific literacy and math skills. Both the IE program and the domain-specific curriculum were taught using the culturally sensitive principles of mediated learning.
- The intensive nature of the intervention. The program format included 5 hours a week of IE, 5 hours of Hebrew, and 5 hours of math.
- Small group format. CoReL was organized for a group of 10–15 students, who were taken out of their classes for 15 hours of CoReL activities.
- Intensive supervision. The application of the CoReL model was closely supervised by experienced IE, mathematics, and reading specialists.

In what follows the results of CoReL groups are presented. The majority of CoReL students were new immigrants from Ethiopia, but the program also included some Israeli-born students of Ethiopian origin as well as immigrants from other regions (parts of the former Soviet Union – Transcaucasia and Central Asia). At the beginning of the program the students were tested by a standard cognitive test – Raven Colored Progressive Matrices. Their results were about one standard deviation below the Israeli age norm. This finding was important, because the original concern of the schools was with mathematics and reading achievement. Teachers and school administration were apparently oblivious to the more general cognitive difficulties experienced by the students. The CoReL program was implemented for 8 months. Table 4.3 shows the results of the pre- and postprogram mathematics tests reflecting the 4th-grade standards that included the concept of number, arithmetic operations with up to three-digit numbers, and simple text problems. An effect size of 1.0 is large. Students also demonstrated significant improvement of their general nonverbal problem solving, measured by the Raven Colored Matrices, and by the end of the program reached the normative Israeli age level. Though we are fully aware that the parallel improvement of general cognitive skills and mathematical achievement does not prove any causal connection, in practice it was impossible not to detect an obvious impact of such general skills as systematic data gathering, analysis of the sample tasks, systematic comparison, and demand for logical justification required in IE on the mathematical task performance.

Table 4.3. Math scores at the beginning and the end of CoReL program; standard deviation in parentheses

Pretest	13.0 (8.4)
Posttest	22.4 (10.4)*
Max score	48
Effect size	1.0

* $p < 0.01$.
$N = 29$.

The studies demonstrated that simple integration of immigrant students into regular classrooms with only regular second-language support is often insufficient for bridging the cross-cultural gap. This is true not only for children who immigrated at the school age but also for younger children who started their formal education in the new country. Educators often perceive the students' difficulties rather narrowly as confined to underachievement in reading comprehension and mathematics. At the same time the results of nonverbal cognitive assessments indicate that at least some, but probably the principal, difficulties stem from the lack of more general cognitive strategies required in all types of academic problem solving. There is at the same time a significant discrepancy between the immigrant students' low level of cognitive performance and their rather strong learning potential that reveals itself under conditions of dynamic assessment that included mediated learning. The IE program seems to be effective in closing the cognitive gap and improving the students' general problem-solving skills. Though the above data are insufficient for determining the casual relationship, the significant enhancement of mathematical performance occurred in parallel with the improvement of the students' general problem-solving strategies. Although in these two studies the "bridging" of IE-based cognitive strategies to mathematics was done in an *ad hoc* manner, the RMT paradigm aims at developing a comprehensive system for providing immigrant and minority students with the general cognitive prerequisites and mathematically specific cognitive tools for coping with mathematical problem solving.

5 Mathematical Concept Formation and Cognitive Tools

Mathematically Specific Psychological Tools

As discussed in Chapter 4, one of the main goals of the Instrumental Enrichment (IE) program is to help learners to appropriate and internalize general psychological tools. This process of appropriation and application guides learners to systemically develop their high-level cognitive functions. However, it is only through the mathematically specific psychological tools that these cognitive functions can be organized and integrated with basic conceptual elements in such a way that they become capable of supporting mathematical generalizations and abstractions.

The prevailing culture of the current mathematics instruction presents learners with ready-made mathematical concepts using the abstract language of mathematical symbols followed by algorithmic deductive demonstrations, examples, and problems that require direct application. This paradigm is used both in the training of preservice teachers in colleges and universities and later for the students of these teachers when they start teaching in various classrooms around the United States. This entrenched model is further promoted by the textbook industry, which ascribes overwhelmingly to presenting mathematics content through the same regimen. As a result a considerable gap emerges between the paradigm of professional mathematicians and that of mathematics education. Professional mathematicians use the existent mathematical symbolic tools or generate new ones to develop, represent, manipulate, and validate mathematical ideas. In a word they are focused on the process of mathematical reasoning and the tools required by this process. On the contrary, the prevailing culture of mathematics education has learners beginning with the "products" of mathematical investigation instead of its process and leads them through a "mechanical" path that has no inherent requirement for either the rigor of mathematical reasoning or the internalization of mathematically specific psychological tools.

Thus, one problem stemming from the prevailing culture of mathematics education is the failure to consider mathematically specific psychological tools as artifacts separate and distinct from mathematical content. These devices are perceived by students as pieces of information or content rather than as "tools" or "instruments" to be used to organize and construct mathematical knowledge and understanding.

Further, mathematical education finds itself in a more difficult position than other disciplines vis-à-vis symbolic tools. On the one hand, the language of mathematical expressions and operations offers probably the greatest collection of potential psychological tools. On the other hand, because in mathematics everything is based on special symbolic language it is difficult for a student, and often also for a teacher, to distinguish between mathematical content and mathematical tools. In Rigorous Mathematical Thinking (RMT) theory, the relationships among tools, inner psychological functions, mathematical activity, and content and their roles in producing mathematical understanding eliminate this dichotomy.

Mathematically specific psychological tools extend Vygotsky's (1979) notion of general psychological tools. Symbolic devices and schemes that have been developed through sociocultural needs to facilitate mathematical activity when internalized become students' inner mathematical psychological tools. Mathematics as we know it today is a cross-cultural synthesis that has evolved through a long, complex infusion of psychological tools and their cultural-historical significance originating from a number of cultures. The structuring of these tools has slowly evolved over periods of time through collective, generalized efforts reflecting the needs of developing cultures. Although each of these tools is distinctively different from the others, in RMT theory they share the characteristics of reflecting the structure/function relationship, contributing to mathematical conceptual understanding and application, and serving to provide the distinct mathematical logic and high levels of precision demanded by and characteristic of the mathematics culture.

The process of appropriating and internalizing a mathematically specific psychological tool based on its structure/function relationship equips the learner with an internalized system of relationships to construct or apply mathematical knowledge at one or more of the three levels described in Chapter 1 – operationally, conceptually, or insightfully. An example of this internalization process is given by Nunes (1999), who cites a literature review by Hatano (1997) about the features of the grandmasters' use of the abacus. The grandmasters acquire an internal spatial representation of the external system of the abacus that is preserved as they perform calculations even while answering verbal questions. It is this conceptualization of mathematically specific psychological tools that the product and the process of mathematical representation can take on greater distinction and meaning for classroom instruction.

In traditional mathematics teaching the external form of mathematically specific symbolic tools are usually called representations. We consider the term *representation* rather unfortunate because it implies a rather passive form of presenting something (e.g., the functional relationship in a certain symbolic form). In our opinion the term *symbolic tools* is more appropriate because they point to the active instrumental function of symbolic artifacts that shape the learner's mathematical reasoning. When this symbolic structure is fully appropriated and internalized by the learner as a mathematically specific psychological tool, the learner acquires and internalizes the relational aspects of its components. At the next stage learners presented with relevant data utilize this internalized system of tools for organizing the data and forming sets of relationships within the data, which, when operated on, leads to a mathematical conceptual understanding. The external structure of the symbolic artifact is mapped with the internalized structure of relationships.

The process of appropriating and internalizing an external mathematical artifact as an internalized tool requires among others the cognitive function *analogical thinking.* An example with *mathematical analogical thinking* is presented in Figure 5.1. The solution of all the problems in Figure 5.1 irrespective of their modality (figural, numerical, or verbal) requires analogical reasoning. In the traditional approach these tasks would be considered a set of mathematical representations. From the RMT perspective, figural, numerical, and verbal symbols must be appropriated and internalized as psychological tools through structural analysis, analogical thinking, and operational analysis. When this occurs in the mathematics education classroom, mathematical symbols move from being mathematical products to becoming mathematical processes. Thus, learners who investigate mathematical symbols do not see a static presentation of data embedded in what could be a tool but construct mathematical knowledge through mathematical reasoning, experiencing real mathematical activity. If learners construct the symbolic representation, they become engaged in designing a method for organizing and forming relationships among mathematical data through mathematical reasoning, thus experiencing a genuine mathematical activity.

The role of each mathematically specific psychological tool in RMT is discussed at length in a different section. In this section we describe some of the most common of these tools.

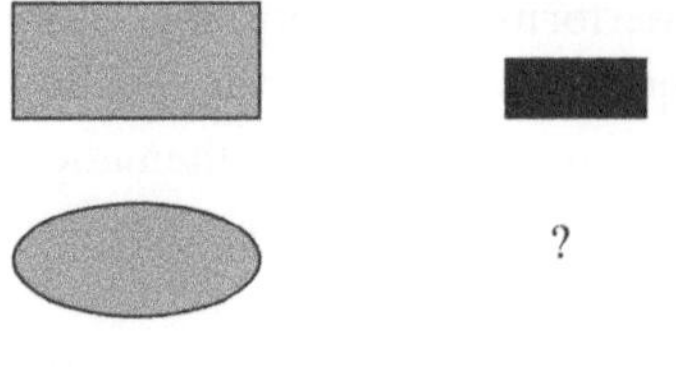

2 5 8 11 14 17
Formula: $X_{n+1} = X_n + 3$

2 6 18 54 ? ?
Formula:

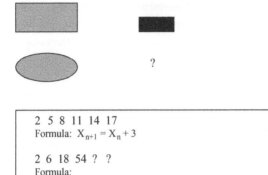

Task 1.
Jim and Tom started walking at the same time using the same path. Jim walked from Wellesley to Brookline with a speed of 3 miles per hour, and Tom from Brookline to Wellesley with a speed of 2 miles per hour. They met in 2 hours. What is the distance between Wellesley and Brookline?

Task 2.
Jim and Tom started walking at the same time using the same path. Jim walked from Marlborough to Greenfield, while Tom walked from Greenfield to Marlborough. Both of them walked with a speed of 3 miles per hour. They met in 2 hours. While they walked a pigeon was flying constantly between Jim and Tom with a speed of 9 miles per hour. How many miles did the pigeon fly before the boys met each other?

Figure 5.1. Analogical reasoning in graphic, numerical, and verbal modalities.

Symbols and Codes

This group of mathematically specific psychological tools consists of three categories of codes and symbols. The first category consists of codes and symbols for forming qualitative relationships, such as the order of operations or geometric relationships (e.g., parallel or perpendicular; see Table 5.1). The second category consists of codes and symbols for encoding quantitative relationships (e.g., $=$, $<$, $>$, and $\gg$) and mathematical operations ($+$, $-$, $\times$, and $\div$). Each of these signifies a defined quantitative operation between the quantitative aspects of two constructs (Table 5.2). The direction of reading from left to right combined with the encoded meaning of the specific process of forming the relationship demonstrates a structure/function relationship.

Table 5.1. Signs and symbols for forming qualitative relationships

Name	Symbol	Sample	Description
Parentheses	( )	$a(6 + c)$	Compute within the parentheses first and then perform the other operation. Add the quantity 6 to the quantity *c* and multiply the product by the quantity *a*.
Point	·	a . . .	A single position in space
Line	↔		An infinite set of points
Line segment	ab		A piece of a line with definite endpoints
Parallel lines	‖	‖	Lines or line segments that are aligned in the same direction and conserve constancy in their distance apart at every point
Perpendicular lines	⊥	⊥	Lines or line segments that intersect to form right angles

Table 5.2. Signs and symbols for forming quantitative relationships

Name	Symbol	Sample	Description
Equal	=	$A = C$	The quantity *A* is equal to the quantity *C*
Greater than	>	$5 > 2 + 1$	The quantity 5 is greater than the quantity of 2 + 1
Less than	<	$B < 10$	The quantity *B* is less than the quantity 10
Greater than or equal to	≥	$D \geq \rho$	The quantity *D* is greater than or equal to the quantity ρ
Less than or equal to	≤	$k \leq 6.78$	The quantity *k* is less than or equal to the quantity 6.78
Addition	+	$7 + 5$	Add the quantity 7 to the quantity 5
Subtraction	–	$12 - 4$	Subtract the quantity 4 from the quantity 12
Multiplication	×	5×8	Multiply the quantity 5 by the quantity 8
Division	÷	$18 \div 3$	Divide the quantity 18 by the quantity 3
Square root	√	$\sqrt{4}$	Take square root of 4

Table 5.3. Signs and symbols for complex and functional relationships

Name	Symbol	Example	Description
Summation	$\sum$	$\sum 25j$	Tool for adding up a series of numerical quantities
Mathematical function of two variables	$y = f(x)$	$y = \cos(x)$	Expresses the functional relationship between a dependent variable and an independent variable
Derivative	d/dx	$d/dx\, e^x = e^x$	A measure of the rate or how one thing changes in comparison to another; a slope
Differential	$df(x) =$	$f(x) = \sin(x - 1)$	Infinitesimal change in the quantity of a variable
Limit	$\lim f(x)\ x \to a$	$\lim(x^2 - 2)\ x \to 0$	Maximum quantity of an item
Integral	$\int$	$\int (2x^3 + 1)dx$ (between 4 and 0)	A specified summation of infinitesimal changes in the quantities of a variable or a functional relationship between two variables
Pi	π	$A = \pi r^2$	Quantitative ratio between the radius and diameter of a circle

Although the origin of some of these modern symbols was rooted in an emerging community of mathematics scholarship, for others it was the workplace of merchants, farmers, and surveyors. For example, the equal sign first appeared in Recorde's algebra, *The Whetstone of Witte*, published in 1557. His rationale for the symbol for equality being composed of a pair of equal parallel line segments was, "bicause noe 2 thynges can be moare equalle." Prior to appearing in mathematical manuscripts, plus and minus symbols were painted on barrels to signify whether the barrels were full (Eves, 1990). In any case the evolutionary processes that led to these modern symbols in mathematics endowed each with a precise sociocultural meaning fitted for the mathematics culture. Cajori (1993) presents comprehensive research on the origin, competition involved, and the spread of mathematical codes and symbols among writers in various countries.

The third category consists of those codes and symbols for forming complex and functional relationships, such as formulae, the $\sum$ sign, derivatives, differentials, integrals, and so on (Table 5.3).

Table 5.4. Base 10 number system

10^6	10^5	10^4	10^3	10^2	10^1	10^0
Millions	Hundred thousands	Ten thousands	Thousands	Hundreds	Tens	Ones

Number System with Its Intrinsic Place Values

RMT theory considers a number system with its intrinsic place values as a mathematically specific psychological tool that organizes and forms precise part/whole relationships regarding quantity. Its structure consists of sequenced horizontal positions in linear space that are used to form a set of progressive part/whole relationships with each other, using powers of the base of the number system. This structure for a segment of the base 10 number system is given in Table 5.4.

Thus, various combinations of the digits for the number system (in this case 0 to 9) can be spatially arranged horizontally to express any magnitude of part/whole relationships with consistency and precision. Although a number can appear in isolation, the meaning of its underlying construct becomes clear only through a system of other constructs, which together form an abstract context of part/whole relationships. This structure of relativity depicts interdependency among the parts that can be used to organize and form relationships among quantities (amounts of some object) as well as sequence items in temporal relationships. For example, the number 4 may appear alone, but its meaning is established only from the collection and ordering of other constructs underlying 1, 2, 3, 5, 6, 7, 8,.... In other words, the conceptualization of 4 implies a conceptualization of 2, whose construct is half the value of 4 and has twice the value of the construct represented by 1, and so on. These quantitative values also dictate their sequencing. Thus, the constructs represented by number symbols help learners create and define new relationships and patterns through a structure/function relationship. When learners have fully appropriated this structure as a tool they acquire an internalized system of relationships that allows them to organize, sequence, form relations, and provide mathematical logical evidence regarding quantity. Learners become fully aware that they are equipped to extend this quantifying-sequencing structure to accommodate more constructs and ordering through its structure/function relationship.

This systemic context of numbers is deeply embedded in the history of mathematical thought. Researchers and writers on the history of mathematics, when presenting numerical artifacts, almost always report on a group of

Figure 5.2. Number line.

symbols that demonstrate a system of relationships (see Cajori, 1993; Eves, 1990; Gillings, 1972; Kline, 1972; Smeltzer, 2003; Smith, 1958). An example comes from the earliest artifact of the use of numbers discovered by de Heinzelin in the Congo (Zaire) (see Eves, 1990; Sertima, 1984; Zaslavsky, 1984). The Ishango bone, dated at 8000 years old, has a system of markings that researchers analyzed to be a number system and appears to be a mathematical tool (Sertima, 1984, p. 14; Zaslavsky in Sertima, 1984, pp. 110–126).

Number Line

The structure of a number line stems from linear space that has been analyzed into equal-sized segments with each segment representing the same range of quantitative value as the others (Figure 5.2). The alignment of these sequenced segments is used to organize quantitative values into sequenced part/whole relationships that are sequentially encoded with numbers. These segments of linear space may be further analyzed into equal-sized parts and encoded appropriately, thus further differentiating the part/whole relationship. When such structures have been fully appropriated by learners as a tool, they can use these internalized sets of relationships to analyze, compare, form proportional relationships, sequence, and provide logical evidence about quantity. The appropriation and internalization of this tool helps learners to understand that each part is a whole while at the same time it is a part of numerous larger wholes, a defining aspect of the construct "number."

Table

A table is one of the most common symbolic tools that can be used both as a general psychological tool (see the section on IE and psychological tools in Chapter 4) and as a mathematically specific tool. It is possible that the origin of the structure of the table is associated with the practical needs of double-entry booking (see Crosby, 1997, pp. 199–223). The structure of every table consists of columns and rows, with each column having a heading that organizes data or information in that column into a category or a set. As learners move in the horizontal direction from the left to the right in one row they are forming a relationship between the item in the first column and the corresponding item in the second column. As this movement continues to the third column the learners form a relationship between the first relationship and the corresponding item in the third column, and so on. Thus, when learners internalize the structure of a table as a tool, they become equipped with internalized abstract relational thinking to organize information into sets, form relationships within the data, and form relationships between relationships. A particular example is when learners are presented with data in tabular form for two variables in a mathematical relation or function. They can apply the internalized mechanism of this tool to construct understanding of the functional relationship between the two variables and how the behavior of one variable affects the behavior of the other variable and derive this functional relationship. Conversely, when learners are presented only with a mathematical equation involving two variables they can utilize this tool to produce, organize, analyze, and form relationships from specific data and further characterize the mathematical relationship.

Cartesian Coordinate System

The appropriation and internalization of mathematically specific psychological tools by learners is proceeding from the structure of the tool to its function. However, the course of action taken by the developers of mathematical tools is just the reverse – from function to structure. The historical development and emergence the Cartesian coordinate system progressed because of the need for a method defined by a prescribed function. Various efforts have been made to design and fabricate a structure that would operationalize the method for analyzing curves and surfaces, as described in Chapter 1.

The structure of a coordinate system consists of two number lines, one drawn horizontally and the other one drawn vertically, which intersect at their origins to form four right angles (see Figure 5.3). This structure really consists of two-dimensional space that has been analyzed into four quadrants and further analyzed into smaller equal-sized two-dimensional squares or rectangles. Because each number line can be used to analyze, compare, and form relationships between quantities, these two number lines together can be used to form relationships between corresponding values for two variables. Internalizing this structure as a tool will equip the learner with two mathematical capacities. One capacity is to organize and form relationships between data for any function of two variables and construct this functional relationship as a graph. The second is to analyze a graphed functional relationship that will lead to constructing a conceptual understanding of the interdependency of the two variables.

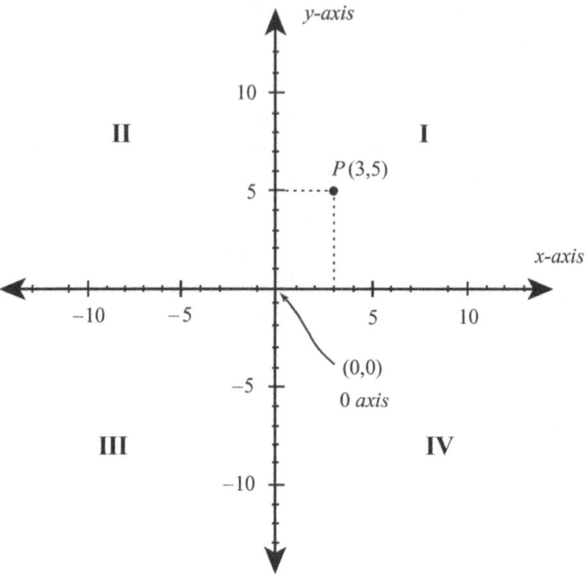

Figure 5.3. Coordinate system.

Mathematical Language

Contemporary literacy research (Olson, 1994; Scribner and Cole, 1981) recognizes that different types of literacy have a different impact on human reasoning and psychological life in general. In a very broad sense literacy shifts human thought from the domain of natural objects and processes to the domain of symbolic objects. It is our contention that advanced mathematical activity has its origin in symbolic thought. Connected to this contention is the question, "What has been the path for the natural historical unfolding between symbolic thought and mathematical expression?" As presented in Chapter 1, the earliest appearance of mathematical activity seems to center on development of numeral systems by the Egyptians and Babylonians, with first evidence of mathematical artifacts, from 3000 B.C. to A.D. 200 (Boyer and Merzbach, 1991; Eves, 1990; Gillings, 1972; Kline, 1972). The history of mathematical symbolization seems to be closely intertwined with the history of literacy.

All mathematical activity must center on generalizations and abstractions about patterns and relationships through chains of logical thought. Such activity can range from investigating and analyzing patterns and relationships to generating and describing generalizations and abstractions to applying and creating new insight from them. The apex of mathematical activity is the formulation of mathematical statements and the validation or invalidation of such statements. Mathematical language is both the medium and process underlying such activity. We define mathematical language as the integrated use of symbols, formulae, mathematically specific psychological tools, and verbal definitions and terms interwoven with propositional logic, precision, and concisely defined rules to articulate mathematical activity. The use of the symbols, tools, and verbal terms according to the rules, such as order of operation, comprise the syntax of mathematical language. The use of propositional logic comprises the semantics of mathematical language that defines its meaning or truth. Thus, mathematical language is both a vehicle for transmitting the culture of mathematics from generation to generation and a process for structuring and expressing mathematical understanding.

In Vygotsky's (1986) theory language is seen both as a symbolic tool shaping other cognitive functions and as one of the higher psychological functions. In this regard, the language of mathematics serves both as a tool for shaping mathematical thought and as one of the higher order mental functions. In the RMT paradigm the instrumentality of the language of mathematics can be viewed from the perspective of how it organizes and transforms the learners' everyday language and spontaneous concepts into more unified, abstract, and symbolic expressions. Mathematical language helps learners to organize and integrate their mathematical thinking and conceptual elements for the formation of mathematical ideas. Instrumentally it supplies learners with the structures and facility to analyze and evaluate mathematical statements and strategies, formulate explanations, generate questions, and construct conjectures and arguments that comply with the rules and logic of the mathematics culture.

The functionality of mathematical language as a cognitive process stems from the fact that this language, as it emerges in learners, progressively becomes the conceptual content of learners' thinking. It also becomes the operational medium for the expression of mathematical thought and conceptual understanding. Expressing mathematical thought is itself a domain-specific cognitive process. Moreover, the transformation of initial mathematical thought into a proper mathematical expression provides both the activity and content of learners' metacognition. Thus, on the one hand, mathematical thought is engendered by the learners' inner speech about their own mathematical thinking. This inner verbal process leads to the elaboration of the relationships among different segments of learners' thought. On the other hand, there is always the "external" mathematical language that is internalized by learners and becomes their tools for translating their mathematical thought into the publicly comprehensible statements. Following Vygotsky's (1986) terminology one can say that the inner "sense" of learners' mathematical thinking should be translated into the written mathematical text that follows the lines of common conceptual "meaning." In the process the inner mathematical thought starts being shaped by the symbolic tools provided by mathematical language. This requires a more intense and expanded engagement of learners' inner "self," as internalized forms of mathematical symbolism become the process and content of written expression. As a result learners see their mathematical intellectual self engaged in mathematical scholarship.

Thus, mathematical language is essential as both a tool and a function in moving learners from symbolic thought through cognitive processing to mathematical expression. It is the operational substance for oral and written expositions about relationships and patterns verbally and symbolically and through the other mathematically specific tools, such as tables, coordinate systems, diagrams, learners' idiosyncratic notations, and schemes. Learners acquire meaning from mathematical language that everyday language cannot produce. Consider the following examples.

Example A: $3x^2 + 2x = 0$
Equivalent expressions in mathematical language:
1. 3?? + 2? is same as zero;
2. 3 times *x* times *x* plus 2 times *x* is equivalent to 0;
3. Take some value of *x* and multiply it by itself and then multiply the product by 3 to get product A, now multiply this same value for *x* by 2 to get product B. Product A must be the negative of product B.

Example B: $y = 2x + 7$
Equivalent expressions in mathematical language:
1. 2 times *x* plus 7 gives the dependent amount of *y*;
2. The value of *y* is the same as the total value of 2 times some value of *x* plus the quantity 7;
3. For every quantity or value of the independent variable, *x*, multiply it by 2 and to that product add the amount 7 to get the corresponding value or quantity of the dependent variable, *y*.

Example C:
1. A learner using everyday language to describe movement along Lakeshore Drive in a car. "My father was driving our red blazer north on Lakeshore Drive with my mother sitting on the passenger's side and

my baby brother and I were sitting on the back seat. Lakeshore Drive twists and turns like the shore of Lake Michigan as we drove from 63rd Street Beach to McCormick Place. Dad drove as fast as he could without driving over the speed limit of 55 mph but he had to slow down several times because of slower drivers in front of us."

2. An advanced learner using mathematical language to describe the same movement: "Some object A moves from point P_1 to point P_2 along a curvilinear path that is parallel to an adjacent curvilinear path at a varying rate of speed that is a function of other variables and with a maximum rate of 55 mph."

Example D:

1. The meaning of words in most languages depends on the order in which they appear.
2. In everyday language, "Flip the light switch and close the door" is not the same as "close the door and flip the light switch."
3. In mathematical language $5 + 7 \times 3$ does not mean add the quantity 5 to the quantity 7 and multiply the result by 3 to get 36. Instead, it means multiply the quantity 7 by the quantity 3 and add to that product 5 to get the quantity 26, because in the mathematics culture we multiply and divide before we add and subtract.

Standards-Based Conceptual Formation Through RMT

The term *standards-based concepts* refers to those academic mathematics and science concepts that are central to the national curriculum guidelines for mathematics and science education formulated through education reform efforts in the United States (see Chapter 2). In addition to mathematics content areas of number sense, algebra, geometry, measurement, and data analysis and probability, NCTM (National Council of Teachers of Mathematics, 2000) included separate standards for what may be considered the more process-oriented areas of problem solving, reasoning and proof, communication, connections, and representation. In a standards-based learning environment, students are at the center of the curriculum-instruction-assessment process. Teaching begins with activating the learners' prior knowledge and experience. What is to be taught or learned is constructed from the individual student's and the class's rich experiential repertoire. The core mathematical concepts and the requisite cognitive functions, however, always serve as a guideline for directing these experiences toward conceptual mathematical understanding.

Table 5.5. Alignment of cognitive functions and core mathematical concepts

Core concept	Codes for related cognitive functions
Quantity	$B_2, B_3, B_4, C_1, C_6, C_7, C_8, C_{10}, C_{12}, C_{13}$
Relationship	$A_1, A_2, A_3, A_4, A_5, B_1, B_2, B_3, B_4, B_5, C_1, C_2, C_3, C_4, C_5, C_6, C_7, C_8, C_9, C_{10}, C_{11}, C_{12}, C_{13}$
Representation	$A_1, A_4, B_2, B_3, B_5, C_1, C_5, C_9, C_{10}, C_{11}, C_{12}, C_{13}$
Generalization-abstraction	$A_1, A_4, A_5, B_2, B_3, B_5, C_4, C_5, C_6, C_7, C_8, C_9, C_{10}, C_{11}, C_{12}, C_{13}$
Precision	$B_2, B_3, B_6, C_3, C_6, C_7, C_8, C_9, C_{11}, C_{12}, C_{13}$
Logic/proof	$C_2, C_4, C_{10}, C_{11}, C_{12}, C_{13}$

Conceptual Nature of Cognitive Functions

As stated earlier, the RMT paradigm defines a cognitive function as a specific thinking action that has three components – a conceptual component, an action component, and a motivational component. The conceptual components of the cognitive functions play a central and systemic role in equipping learners with the facility for constructing mathematical knowledge. We have identified six core mathematical concepts that are essential for defining the nature of the mathematics culture. These core mathematical concepts are foundational to mathematical activity and for making mathematical sense. The essential characteristic of mathematical notions as concepts derives from their systemic nature and their orientation toward cognitive processes of the learner. The meaning making in mathematical learning depends on the core concept serving as the subject or the objective of carrying out the function. In this regard the natural alignment of the related cognitive functions (see Chapter 4, Table 4.1) and the respective core mathematical concepts is shown in Table 5.5.

The first core concept is quantity. Quantity answers the question, "How much?" and means amount, value, or magnitude. Patterns and relationships exist all around us, but an important element that allows these patterns and relationships to take on mathematical character is the fact that they are quantifiable. From basic through advanced mathematics, the idea of number is both central and intrinsic to the nature of the subject. However, in RMT theory and practice, the big idea of quantity emerges through relativity of part/whole relationships. The idea of number, or, more precisely, a number system with its place values, is a mathematically specific psychological tool for helping the learner to organize, sequence, form relations, and provide mathematical logical evidence regarding quantity (see previous section on mathematically specific psychological tools). Basically, the notion of quantity is inherent in the objective of the cognitive function *analyzing-integrating*. When a whole (something complete in its amount) or a quantity is analyzed it is broken into smaller parts or amounts and the whole has a larger magnitude or amount than each of the parts. When the learner integrates parts to form a whole he or she is merging amounts of some object or concept to compose or constitute a more complete object or concept. Thus, the conceptual component of *analyzing-integrating* builds the meaning of quantity of some object or concept through parts-to-whole analysis.

An example of a cognitive function whose meaning and objective center mostly on quantity is *quantifying space and spatial relationships*. When exploring the amount or quantity of space or spatial relationships the learner must turn to some psychological tools to deal with such an abstract notion. The most basic of such tools that constitute the learner's internal reference system is his or her own body. This tool is stable (left is always opposite to right and front to back) and when internalized helps the learner to develop spatial orientation and navigate through space by articulating, analyzing, and forming spatial relationships. A more abstractive and specialized psychological tool that helps to further develop the learner's spatial orientation is an external reference system (compass points) that is stable and absolute and can be interfaced with the learner's internal reference system. However, these general psychological tools are insufficient for a quantitative approach to space. To form quantitative spatial relationships, the learner must turn to mathematically specific psychological tools such as the number line for linear space, the *x*-*y* coordinate plane for two-dimensional space, and the *x*-*y*-*z* coordinate system for three-dimensional space.

Basic mathematics skills in RMT consist of the use of certain groups of cognitive functions to deal with the concept of quantity while complying with the mathematics cultural needs for relationship, representation, precision, logic, and abstraction. RMT builds student facility and skills regarding operations and properties through building conceptual understanding of quantity, relationship, and logical processes dealing with quantities. At the same time RMT develops both the need and the ability to use various cognitive functions through such conceptual understanding. In this way the cognitive processes of learners are constantly connected with the conceptual aspect of mathematical knowledge. In addition, it is partly through the notion of quantity that basic mathematics is connected to algebraic thinking and higher mathematics. For example, from a dynamic perspective a *variable* is something that changes its

Rigorous Engagement for RMT

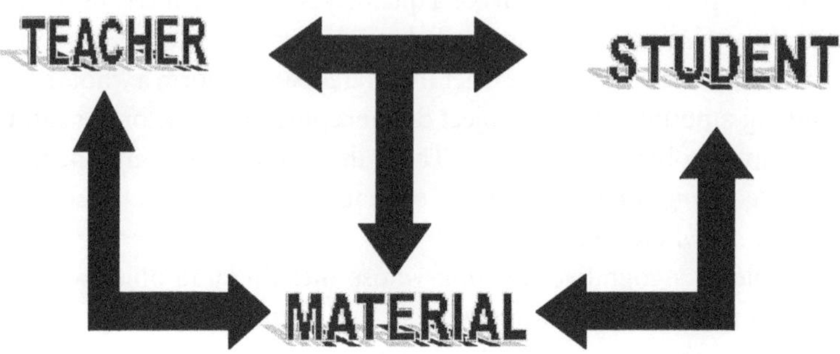

(External Environment)
Lesson/Content

The interactions developed through rigor are dynamic (exciting, challenging, and invigorating), interdependent, and transformative. When these bidirectional interactions permeate each other to produce dynamic reversibility throughout the channels of interaction, rigorous engagement has been initiated.

Figure 5.4. Rigorous engagement for RMT.

value or quantity at different times or in different situations. All functional relationships between variables must address the idea of quantity as one basis of their critical attributes. The basic mathematical operations and properties provide the logical mechanisms for treating quantity in the mathematics culture from arithmetic to advanced mathematics and science.

We discuss other aspects of the natural alignment of the related cognitive functions and the respective core concepts in the next section on the construction of specific mathematics concepts through RMT.

The RMT Process for Concept Formation

The application of RMT focuses on mediating the learner in constructing robust cognitive processes while concomitantly building mathematical concepts using the three-phase, six-step process described next. The process does not take place in a lockstep linear fashion. However, each one of the phases and the steps is essential for learner's engagement in mathematical conceptual understanding. We emphasize that RMT teaching is a process that transforms mathematics content through rigorous engagement (see Figure 5.4).

As placement of a tea bag in hot water is followed by diffusion of the pigments and the experience of the richness of the flavor, aroma, and nutrients, RMT engagement involves cognitive, affective, and conceptual dimensions. In this sense, the RMT process is an infusion that energizes and expands the learning of mathematics conceptual development and problem solving.

Phase I: Cognitive Development

1. The learner is mediated to appropriate the models in the cognitive tasks as general psychological tools based on their structure/function relationship.
2. The learner is mediated to perform the cognitive tasks through the use of the psychological tools to construct higher order cognitive processes.

Phase II: Content as Process Development

1. The learner is mediated to systemically build basic essential concepts needed in mathematics from everyday experiences and language.
2. The learner is mediated to discover and formulate the mathematical patterns and relationships in the cognitive exercises.
3. The learner is mediated to appropriate mathematically specific psychological tools (i.e., number system with place values, number line, table, *x-y* coordinate plane, and mathematical language), based on their unique structure/function relationships.

Phase III: Cognitive Conceptual Construction Practice

1. The learner is meditated to practice the use of each mathematically specific psychological tool to organize and orchestrate the use of cognitive functions to construct mathematical conceptual understanding.

During the entire process, the rigorous nature of conceptual reasoning is continuously evoked and maintained. The RMT theory defines mathematical rigor as that quality of thought that reveals itself when learners are mediated to be in a state of high vigilance – driven by a strong, persistent, and inflexible desire to know and deeply understand. When this rigor is achieved, the learner is capable of functioning both in the immediate proximity as well as at some distance from the direct experience of the world, and has an insight into the learning process. Such a state is directly related to a high level of metacognitive activity with learners constantly reflecting on their own and others' cognitive processes. This quality of engagement compels intellectual diligence, critical inquiry, and intense searching for truth – addressing the deep need to know and understand.

Rigorous mathematical thinking in the learner is characterized by two major components: (1) Disposition of a rigorous thinker – being relentless in the face of challenge and complexity and having the motivation and self-discipline to persevere through a goal-oriented struggle. It also requires an intensive and active mental engagement that dynamically seeks to create and sustain a higher quality of thought. (2) The qualities of a rigorous thinker – initiated and cultivated through mental processes that engender and perpetuate the need for the engagement in thinking. The qualities of a rigorous thinker are dynamic in nature and include a sharpness in focus and perception; clarity and completeness in definition, conceptualization, and delineation of critical attributes; precision and accuracy; and depth of comprehension and understanding. As shown by Schmittau (2004) one of the major challenges facing the U.S. elementary school students exposed to Vygotskian mathematics curriculum is the necessity to sustain concentration and intense focus required by this program. However, on the completion of this curriculum, they were able to solve problems normally given to U.S. high school students.

The RMT process demonstrates the theory of RMT in practice. The three phases and the six steps characterize a quality of classroom instruction that seeks to engage all learners in thinking about thinking to construct mathematical conceptual learning with deep understanding. As mentioned above, the phases and steps do not take place in a rigid, linear fashion but should be used over the course of planning and instructional engagement designed to teach a particular mathematical concept.

Strategic Planning to Teach Through the RMT Process

Implementing the RMT process delineated earlier requires deliberate strategic planning in advance by the teacher. This planning will be most effective through collaboration with other RMT practitioners. The planning should be thorough and demands that the planner(s) should be metacognitive in considering both the process and content of the planned instruction. Each RMT lesson is to be taught and learned through mathematical activity. Such activity is designed to both appropriate methods and tools for conceptual learning and construct cognitive processing in the learner. Through the RMT process the teacher and learner will structure rigorous engagement with mathematical learning activity, thus transforming mathematical content into actions that (a) create a structural change in the learner's understanding of mathematical knowledge, (b) produce systemic conceptual formation, and (c) equip the learner with the language and rules of mathematics (see Figure 5.4).

The first requirement in the planning process involves a structural analysis of the targeted mathematical concept or the big idea and organizing the resulting conceptual components from the most basic to the most complex. Such organization provides the instructional pathway for scaffolding student engagement in the mathematics content through the criteria of mediated learning experience (MLE; see Chapter 4). Next, the teacher must identify the most important cognitive functions needed to equip learners with building their understanding of the components of this conceptual pathway. When a judicious selection of the cognitive functions required to construct the conceptual components of the big idea is mediated during the learning engagement, the cognitive development phase described above will bring about an emergence of the content as process phase, primarily through the natural conceptual nature of the selected cognitive functions.

In the next phase of planning the teacher must choose relevant cognitive tasks with the accompanying symbolic artifacts that are to be appropriated as general psychological tools needed for the cognitive function development. The teacher has to think about and plan how to apply the MLE criteria of *intentionality/reciprocity, meaning,* and *transcendence* to help learners clearly perceive the structure of each external artifact and internalize the relationships of its components to carry out the function of the tool to perform the cognitive tasks. The teacher has to plan how to apply the criteria of MLE to assist the learners to explicitly identify, define, and describe how they are using each cognitive function that will emerge as the cognitive tasks are being performed.

The teacher must now plan to identify the core concept or concepts that will be central to understanding the big idea and develop a mediation strategy for building the core concept(s) by tapping into the learners' everyday concepts and language. While the core concept(s) is/are emerging the teacher has to figure out how to guide learners to explore, discover, and formulate the mathematical patterns and relationships most relevant to the big idea in the cognitive exercises. This aspect of the planning process demands that the teacher apply metacognition to go beyond the ordinary and use ingenuity, expertise, and mathematical knowledge gained from study and prior mathematical education experiences.

In addition, the teacher has to figure out how to guide learners toward appropriation of other relevant mathematically specific psychological tools and design rich opportunities for them to practice the use of cognitive functions and tools for forming and internalizing the targeted mathematical concept.

In the sections that follow we present how the RMT process brings about mathematical conceptual formation by using vignettes, discussions, student and teacher artifacts, and illustrations.

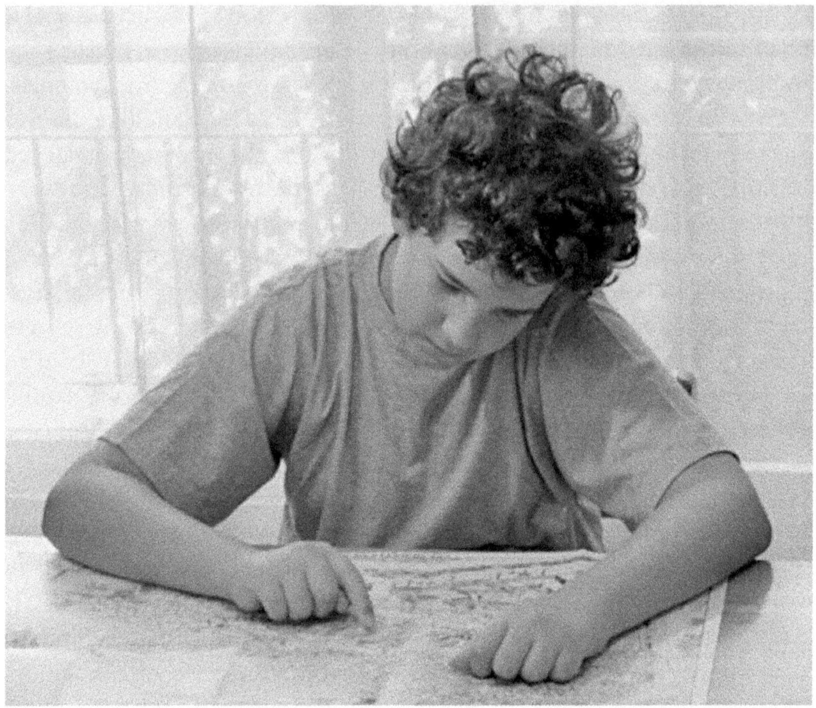

Figure 5.5. Child with a map. (Photo by G. Vinitsky. Used with permission.)

Conceptual Formation in Basic Mathematics Through RMT

Basic Mathematics Operations

As stated in the previous section, one of our premises is that the learners' difficulty in developing mathematical conceptual understanding is partly due to the lack of cognitive prerequisites rather than their ability to gain mathematical knowledge. The RMT-oriented teacher is required to recognize elements of cognitive processes displayed by the students and then discuss an analysis of each of the cognitive functions. In this way students acquire knowledge with analytic cognitive tools and cognitive terminology. Instead of starting with mathematical problems the teacher may start with a picture that is closer to the everyday life of his or her students (see Figure 5.5). The following vignette shows how a simple picture can be used for the initiation of cognitive development.

TEACHER: What is the boy doing?

STUDENT 1: He looks like he is trying to figure out how get to where he wants to go.

STUDENT 2: Yeah. It looks like he is reading a map.

STUDENT 3: He must be thinking of a plan for seeing the different animals at the zoo.

TEACHER: I heard you say that he is thinking of a plan. Do you have to think to figure out where you want to go?

STUDENTS: Yes!!

TEACHER: Do you have to think to read a map?

STUDENT 2 AND SOME OTHER STUDENTS: Yes!!

TEACHER: So we all agree that the boy is thinking. Here is a word that means thinking. I will spell it for you. C-o-g-n-i-t-i-v-e. Class, try to pronounce this word.

[Teacher has students practice pronouncing "cognitive."]

TEACHER: Suppose someone comes through the door. What would you have to do to know if you recognize this person?

STUDENT 1: Look at the person and try to remember in my mind if I have seen this person before.

STUDENT 4: Yeah. I would have to look at the person and think, "try to form a picture in my mind to see if I know this person."

TEACHER: That's very good. You are both telling me you have to think in order to know if you recognize this person.

[Students agree verbally and by giving gestures of approval.]

TEACHER: Let us spell the word "recognize."

[Students spell the word while the teacher writes it on the board.]

STUDENT 5: I see part of cognitive in the word recognize.

STUDENT 6: To recognize someone you have to think.

STUDENT 5: You have to think also to know that you don't recognize the person.

TEACHER: Great!!! So "cognitive" means "thinking." Here is a word I want you to put with the word cognitive. I will spell it for you. F-u-n-c-t-i-o-n. Pronounce this word.

[The class practices pronouncing "function."]

TEACHER: What is a function?

STUDENT 4: It's how something works.

TEACHER: Good. Give me one word that means "function."

STUDENT 2: Process.

STUDENT 6: Doing something.

STUDENT 7: Movement.

TEACHER: Very good. There is a word that means process, movement.

STUDENT 8: Operation.

TEACHER: Great. There is another word that means operation, movement, process. This word begins with the letter *a*.

SEVERAL STUDENTS IN UNISON: Action!

TEACHER: That's correct. Function means action. So what is a "cognitive function"?

SEVERAL STUDENTS: Thinking action.

TEACHER: Open your Rigorous Mathematical Thinking journals and turn to the first page that you have headed "table of contents." The first item in your "table of contents" is "cognitive functions." Write this term and show that it appears on page 2 in your journals. [Pause.] Now turn to page 2 and write "cognitive functions" as your heading. [Pause.] In parentheses under this heading write the meaning of "cognitive functions." What goes in the parentheses?

STUDENTS IN UNISON: Thinking actions!!!

TEACHER: Let us focus again on the picture of the boy at the zoo.

[Teacher pauses to intentionally get students to look carefully at the picture.]

TEACHER: You are going to start using a cognitive function as soon as I ask you this question. We all agreed that the boy is thinking, correct?

[Students agree.]

TEACHER: What tells you that the boy is thinking?

STUDENT 7: It looks like he is focusing.

STUDENT 3: He is pointing to something on the map.

STUDENT 8: It looks like he is trying to figure how to get from where he is to where he wants to go.

TEACHER: This is real good. What name might you give to all these things you are giving me to convince me that you know that the boy is thinking?

STUDENT 1: Clues!

TEACHER: Good! What is another word for clues?

STUDENT 9: Hints!

TEACHER: Very good! What are these clues and hints providing that will help someone know that the boy is thinking?

STUDENT 10: Proof!

TEACHER: That's wonderful! Now give me another word that means proof?

[Pause.]

STUDENT 11: Evidence!!

TEACHER: Very good! Now what if the boy was blowing bubbles and someone claims that the boy is thinking?

STUDENT 5: That would not make sense because it will not take any serious thinking to blow bubbles.

STUDENT 12: Yeah. The proof has to make sense.

TEACHER: You were providing logical evidence to convince me that the boy is thinking. Write this down as your first cognitive function.

The teacher writes "providing logical evidence" on the board and directs students to practice pronouncing it. The teacher then scaffolds students to recognize that they have produced the following meaning of this cognitive function during the previous classroom conversation with the teacher: "Giving supporting details and clues that make sense and serve as evidence and proof for a claim, a hypothesis, or an idea." The teacher mediates transcendence by guiding students to identify examples of logical evidence a physician, a meteorologist, a detective, and a lawyer have to provide to be effective when doing their work.

RMT engagement of students always provides the teacher with various options to address mathematical content through cognition. The teacher, for example, might mediate transcendence to a more mathematically specific context by asking students the question "What is the answer to sixty-eight plus twenty-five?" Once the answer, 93, is given the teacher should require students to collaborate in small groups to provide all of the logical evidence that this is a mathematically true statement. To respond to this requirement, students will have to activate their prior mathematical knowledge regarding the core concepts of quantity, relationship, and logic. Although students often possess considerable mathematical knowledge, much of this knowledge is fragmented and does not appear as a system. Some students' responses will undoubtedly provide the teacher with the opportunity to guide the students toward the base 10 number system, with its place values as a mathematically specific psychological tool through its structure/function relationship.

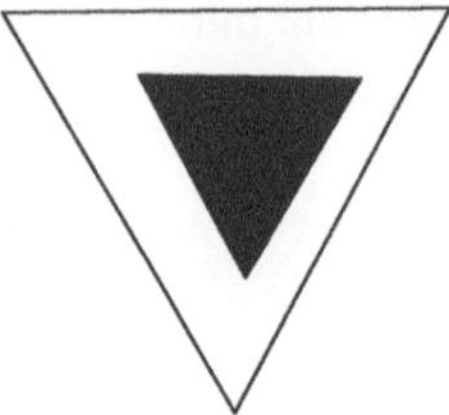

Figure 5.6. Compare the triangles.

What is needed, however, is to separate the number system with its place values from the incidental content and guide the learners to explore and study it as a tool that transcends a particular case. As students internalize the structure in the relationships of place values they will begin to appreciate the unlimited role of number as a tool for providing mathematical logical evidence regarding quantity in all of mathematics. Numerous other examples, as mathematical statements or propositions, might be presented to learners that will require them to *activate prior mathematical knowledge* to *provide mathematical logical evidence* regarding the truth of such statements or propositions. Such practice is an ongoing theme in RMT teaching.

One of the central operations in mathematics and other disciplines is conceptual comparison. When the two objects presented in Figure 5.6 are compared conceptually, learners are guided to produce the data given in Table 5.6. The mechanism of the comparing process begins by choosing and, if necessary, defining the concept by which to compare followed by categorizing and labeling each object separately according to the chosen concept and then comparing the outcomes from the categorizing and labeling to determine if they are similar or different. Of course, the learner has to start building or activating and using the following cognitive functions: *labeling-visualizing, searching systematically to gather clear and complete information, using more than one source of information at a time, encoding-decoding,* and *conserving constancy*. Without this deliberate strategy and cognitive functions the quality of the comparing will be deficient. For example, notice that the object on the left is similar to the object on the right when comparing by the concept of color but is different when comparing by the concept of color location. Color location is formed by combining two concepts, color and location. Both color and location of the color must be considered by the learner, which is more complex than comparing by a single concept, such as color. Comparing, without going through the strategic process delineated above, may produce a different outcome, such as one that results from confusing the two concepts, color and color location.

Table 5.6. Conceptual comparison

Concept by which to compare	Object on left	Object on right	Similar	Different
Figure	Triangle	Triangle	x	
Orientation	Up	Down		x
Color	Black and white	Black and white	x	
Color location	White inside black	Black inside white		x
Color proportion	?	?		

Another possible basis for comparison is color proportion or the amounts of colors as they relate to each other. This demands that the learner uses the following additional cognitive functions: *quantifying space and spatial relationships, being precise, projecting and restructuring relationships*, and *forming proportional quantitative relationships.* It is important to note that there is an optical illusion of size differentiation of the objects when observing the quantity of the color composition of the item on the left with the item on the right. Once there is the recognition that the optical illusion of size does exist and that the large and small triangles on the left are equal in size, respectively, to the large and small triangles on the right an examination of color quantity can be made. First, it appears that it is difficult or impossible to know which of the two colors has the larger amount for each object, white or black. It is this predicament that offers the opportunity for engaging learners in richer mathematical activity. The teacher can present the learners with a number of mathematical propositions. To overcome the optical illusion the first proposition is as follows: the large triangle on the left is equal in size to the large triangle on the right and the small triangle on the left is equal in size to the small triangle on the right. The second proposition is as follows: the base and height of each triangle are known and are labeled b_{lb} and h_{lb}, respectively, for the large black triangle and b_{sw} and h_{sw}, respectively, for the small white triangle. This convention is also used to determine the base and height of the triangles on the right. Thus, the amount of white for the object on the left is $\frac{1}{2}b_{sw} \times h_{sw}$, which must be compared to $\frac{1}{2}b_{lb} \times h_{lb} - \frac{1}{2}b_{sw} \times h_{sw}$ to determine the quantity of white to the quantity of black. Similarly, the amount of black for the object on the right is $\frac{1}{2}b_{sb} \times h_{sb}$, which must be compared with $\frac{1}{2}b_{lw} \times h_{lw} - \frac{1}{2}b_{sb} \times h_{sb}$ to determine the quantity of black to the quantity of white. Because the large black triangle on the left is equal in size to the large white triangle on the right and the size of the small white triangle on the left is equal to that of the small black triangle on the right, the color proportion of black to white on the left will be equivalent to but opposite the color proportion of black to white on the right. Of course, such conclusion to some may appear to be obvious, but to guide learners to engage in the theoretical propositional conjecturing involves them in mathematical activity using reasoning and mathematically specific psychological tools that will be absent by simply concluding the obvious.

Basic mathematics is mostly anchored in the core concepts of quantity, relationship, logic/proof, representation, and precision. In RMT, the learner is guided to deal with quantity through the measurement of objects. The measurement presupposes taking into account which concept or dimension is used and then *comparing* parts to parts and parts to whole and *forming proportional relationships* between and among them. A number system with its unified relativity of parts to whole is needed as a tool to produce and maintain the logic and precision in representing elements in this structure of abstraction. The learner is mediated to perform a series of cognitive tasks presented in different modalities on ordering or sequencing size relationships of objects using the symbols > and <. The first step required of the learner is to identify the most logical and meaningful concept or dimension on which to base the ordering or sequencing. Thus, the prerequisite set of cognitive tasks presented above is needed to teach the learner to *compare* based on the superordinate concept to advance beyond comparing by using individual attributes.

We describe now mathematical activity in RMT that engages the learners with the concept of quantity from a more theoretical perspective. The cognitive function *analyzing-integrating* provides the underlying mental actions for mathematical operations. Learners are introduced to this cognitive function through the illustration shown in Figure 5.7. Here the large square is *analyzed* into four smaller squares when one takes the perspective of projecting out of the page. Viewing the objects from the opposite perspective the process of *integrating* occurs as the four smaller squares are merged together to reconstruct the whole larger square. In RMT, adding and subtracting quantities are complementary actions that do not take place isolated or separated from each other. Adding requires the underlying cognitive process of *integrating*, while subtracting demands the underlying cognitive process of *analyzing*. To gain the full benefit of *analyzing* the learner must *integrate* to reconstruct the object that is being *analyzed* and vice versa. Thus, for the learners to fully understand what they are doing when subtracting one quantity from another it is necessary for them to consider adding the two quantities.

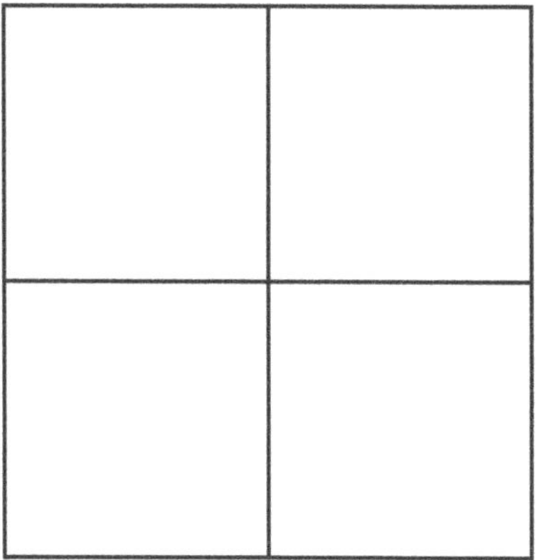

Figure 5.7. Analyzing-integrating.

When adding, the learner forms a linear integration of quantities. This means that the learner merges two or more quantities of the same concept or dimension to compose a new quantity of the same concept or dimension. Although the integrated quantities are in a fixed proportional relationship in the composed quantity, this composed quantity consists of a multitude of other fixed proportional relationships of quantities. In this sense the integrated quantities lose their individual identity when merging to form the quantitative composition that is conceptually homogeneous in nature. Although in mathematics at the most general level we state, for example, that $7 + 5 = 12$ it is true only if 7, 5, and 12 are of the same kind, that is, of the same concept or dimension. For example, 7 pounds plus 5 hours would give us neither 12 pounds nor 12 hours. In addition, the resultant quantity is more than just a combination of two quantities. Although it is true that 7 composes 7/12 of 12 and 5 composes 5/12 of 12, 7 and 5 are not the only quantities that compose 12. The resultant quantity 12 consists of many combinations of other quantities other than 7 and 5. The integration of 7 and 5 to form 12 is somewhat analogous to the molecules of two elements undergoing a chemical reaction, such as hydrogen reacting with oxygen to form water. Water is not simply a mixture of hydrogen and oxygen and 12 is not just a grouping of 7 and 5. Whereas hydrogen is a gas at room temperature and will burn and oxygen is also a gas at room temperature and will support combustion, water is a liquid at room temperature and will neither burn nor support combustion. In a similar manner, the quantities 7 and 5 lose their identity when integrated to form 12.

When subtracting, the learner analyzes a quantity in a linear fashion. This means that the learner breaks down a quantity of a concept or dimension into two components of the same concept or dimension when one component is specified. For example, 8 – 3 means that quantity 8 is to be broken down into two parts and one of the parts is the quantity 3. The quantities 8 and 3 along with the missing part 5 must be of the same concept or dimension.

The cognitive function *analyzing-integrating* also underlies the complementary operations multiplying-dividing. When multiplying, the learner is forming a nonlinear integration of quantities. This means that the learner first forms a set relationship of equivalent quantities into a concept or a dimension and integrates a specified number of these sets of quantity relationships. This is similar to Schmittau's (2003) description of multiplication as an action that requires a change in unit. Davydov (1992) describes multiplication as a change in the system of units. In this example, 8 × 6, we form a set of 8 items and we integrate 6 sets of these 8 items to compose the quantity 48. If 8 and 6 had units or dimensions we would be forming a nonlinear unit quantity relationship by connecting a set of equivalent quantities of a concept or dimension to a quantity over another concept or dimension to produce a conceptual or dimensional change and then integrating a specified number of these nonlinear unit quantity concept relationships to form a new quantity of the unit concept or dimension.

When the learners divide they break down a quantity into two nonlinear elements. They are breaking down a quantity with a concept or dimension into two quantity components where one quantity is specified by its concept or dimension as a unit concept relationship and finding the number of sets of this quantity needed to integrate to make the given quantity. After introducing the learners to the mathematical operations, they are given tasks similar to the ones described below where they have to both follow instructions and engage in construction.

Mathematical Activity A on Quantity

1. Draw a short horizontal line segment and *encode* it as E.
2. Draw another horizontal line segment that is twice the length of E and *encode* it as A.
3. Draw a third horizontal line segment that is three times the length of E and *encode* it as C.
4. Write the quantitative relationship between E and A in as many ways as possible using the = sign each time.
5. Write the quantitative relationship between E and C in as many ways as possible using the = sign each time.
6. Write the quantitative relationship between C and A in as many ways as possible using the = sign.

7. Write the quantitative relationship among E, A, and C in as many ways as possible using the = sign.

Here learners must translate verbal instructions, construct quantitative representations of linear space, and construct representations of quantitative symbolic relationships. Overall, learners are required to utilize many cognitive functions, including *encoding-decoding, defining the problem, projecting and restructuring relationships, analyzing-integrating, inferential-hypothetical thinking,* and *forming proportional quantitative relationships.* Guiding learners to perform these tasks may require rigorous mediation. Learners may have to turn to constructing and using their idiosyncratic psychological tools as intermediaries in the problem-solving process. The teacher should plan rich scaffolding to guide learners from the given verbal instructions and relationships to expression of mathematical symbolic relationships. Mediation through these tasks will develop broad zones of proximal development for the learners who will be required to stretch beyond their current performance level.

In these tasks quantity must be viewed through both whole numbers and fractions. For example, results for task 4 in Activity A are $A = 2E$ and $E = \frac{1}{2}A$. Results for task 5 are $E = 1/3\ C$ and $C = 3E$. Results for task 6 require the use of a fraction and a mixed number: $A = 2/3\ C$ and $C = 1\frac{1}{2}A$. Results for task 7 require the complementary actions of adding-subtracting: $C = A + E$, $A = C - E$, and $E = C - A$.

A more challenging set of tasks is as follows.

Mathematical Activity B on Quantity

1. Draw a short horizontal line segment and *encode* it as A.
2. Draw another horizontal line segment that is three times the length of A and *encode* it as B.
3. Draw a third horizontal line segment that is ½ the length of A and *encode* it as C.
4. Write the quantitative relationship between A and B in as many ways as possible using the = sign each time.
5. Write the quantitative relationship between A and C in as many ways as possible using the = sign each time.
6. Write the quantitative relationship between B and C in as many ways as possible using the = sign.
7. Write the quantitative relationship among A, B, and C in as many ways as possible using the = sign.

Understanding and Comparing Fractions with Unlike Denominators

Learners from elementary school through college have great difficulty conceptually understanding comparing and adding fractions with unlike denominators. We described the RMT process as involving the following three phases: cognitive development, content as process, and practice of cognitive conceptual construction. Our structural analysis of the conceptual development regarding fractions with unlike denominators reveals that the learner is required to activate and utilize almost all of level A and B cognitive functions (see Table 4.1 and Table 5.5) along with *activating prior mathematically related knowledge, providing and articulating mathematical evidence, defining the problem, inferential-hypothetical thinking, projecting and restructuring relationships, forming proportional quantitative relationships, mathematical inductive-deductive thinking, mathematical analogical thinking, mathematical syllogistic thinking, mathematical transitive relational thinking,* and *elaborating mathematical activity through cognitive categories.* This cognitive demand alone creates problems for learners who are taught to deal with fractions with unlike denominators algorithmically without being explicitly engaged in cognitive development. Once these cognitive functions are in the process of being developed or are fully internalized by learners, teaching fractions with unlike denominators is greatly facilitated.

Below is the outline of RMT teaching steps leading to conceptual comprehension of fractions:

A. Part/whole relationship.
B. Defining the whole: Establishing the logical basis for doing fractions.
C. What the denominator tells us in relationship to the whole.
D. What the numerator tells us in relationship to the whole.
E. Forming a relationship between the meaning of the numerator and the meaning of the denominator.
F. Proportional relationship reasoning.
G. Constructing text to express the meaning of a fraction.

The following vignette demonstrates how learners with emerging or developed cognitive functions can be guided and mediated to deal with fractions with unlike denominators.

MEDIATOR: Now let's go back a little. Tell me, what do you mean about the whole?

STUDENT 1: You know the whole thing.

MEDIATOR: But what does "whole" mean?

STUDENT 13: The complete thing.

MEDIATOR: Good. Is there another word that means whole?

STUDENT 9: All of it with nothing missing.

STUDENT 11: The entire thing.

STUDENT 4: The full thing.

MEDIATOR: Very good! Today we begin our unit on "fractions." Before we can begin to do anything meaningful with a fraction, we must use the cognitive function *defining the problem.* The very first thing we must do is *define the whole clearly and precisely.* Write this in your notes (teacher dictates): When I have to deal with a fraction, my very first step is to ask myself, "What is the whole?" I must figure out what the whole is and state, using precise language, what the whole is without giving numbers.

MEDIATOR: Now let's practice this first step. We will work in small groups. [Teacher designates the students that will be in each small group.] Remember; don't try to solve the problem. Define the whole and write out your results using complete sentences. Do not use numbers.

Examples:

(1) What fraction of days in the week begins with the letter *T*?
What is the whole?

(2) What fraction of the class consists of female students?
What is the whole?

(3) Mary spent five years in elementary school, three years in middle school, four years in high school, four years in college, and three years in law school. She just had her thirty-fifth birthday after practicing law for about ten years. What fraction of her life did she spend in high school?

MEDIATOR: What is the whole?

[Mediator has small groups to present their results to the class and discuss them. Mediator mediates students' responses to correct misconceptions.]

MEDIATOR: Carla, go to the board and write a fraction, any fraction.

[Carla writes 5/9 on the board.]

MEDIATOR: Now are we told what the whole is?

STUDENT 15: No. Nothing is given about the whole.

MEDIATOR: This is correct. The fraction, in and of itself, does not tell you to think about the whole. When we are given a fraction, we must automatically pause and tell ourselves, "I have to think that there is some whole first. I may not know exactly what the whole is, but I must tell myself that there is something complete that I'm starting with." How do we *label* the number below the line?

STUDENT 6: It's the denominator.

MEDIATOR: What does the denominator tell us?

STUDENT 1: It tells us something about the parts.

STUDENT 12: I think it tells us something that we have to have the same number when we add or subtract fractions.

MEDIATOR: We can't go to that part yet. Before we start doing anything with the fraction we have to have a very clear understanding what the fraction means. That means that we have to understand what each part means first. We said that we must hold in our minds the fact that we have a whole thing or some complete thing. Everything we deal with regarding this fraction is going to have to connect with this idea of some complete thing. Now the number under the line is *labeled* the denominator as we said. The denominator tells us how many equal-sized parts the whole has been *analyzed* into. This is very important. Write in your notes what the denominator tells us.

MEDIATOR: Let's look at Carla's fraction. Let's go back to one of the group's problems that we discussed.

STUDENT 9: The nine tells us that the whole thing is broken into nine equal-sized parts.

MEDIATOR: Very good. It is very important that the parts are equal in size. What is another word for size?

STUDENT 5: Amount.

MEDIATOR: Good! Give me another word.

STUDENT 7: Quantity.

STUDENT 4: Value.

MEDIATOR: Great! So the denominator tells us how many parts of equal size, of equal amount, of equal quantity, or of equal value the whole has been broken down to. What do we call the number above the line?

STUDENT 12: It's called the numerator.

MEDIATOR: What does the numerator tell us?

[Pause.]

MEDIATOR: The numerator connects part of the meaning of the denominator and the whole. The numerator tells us the number of these equal-sized parts we are considering at this time. Take your time and think. Give me a complete and precise answer. What does the fraction five-ninths mean?

[Pause.]

STUDENT 15: This fraction means that a whole something is analyzed into nine equal-sized parts and I am considering five of these equal-sized parts at this time.

MEDIATOR: If you were considering the whole thing, what would you be considering as a fraction?

STUDENT 15: I would be considering nine-ninths.

MEDIATOR: What fraction are you not considering?

STUDENT 15: I'm not considering four-ninths or four of these equal-sized parts.

MEDIATOR: Very good! Let's go back to two of the problems that you worked on in groups.

[Mediator puts this problem on the overhead: What fraction of days in the week begins with the letter *T*?]

MEDIATOR: Now, give me precise meaning for this fraction.

STUDENT 11: The whole is the total number of days in the week. The fraction for the whole is seven-sevenths. The week is *analyzed into seven equal-sized parts or seven days.* We need to consider two of these equal-sized parts because two days of the week begin with the letter *T*. I am not interested in five of these days or the fraction five-sevenths.

MEDIATOR: Great! Now, what about this fraction?

[Mediator puts this problem on the overhead: Mary spent five years in elementary school, three years in middle school, four years in high school, four years in college, and three years in law school. She just had her thirty-fifth birthday after practicing law for about ten years. What fraction of her life did she spend in high school?]

STUDENT 5: The whole is the total amount of time she has lived up to this time, which is thirty-five years. Her life is *analyzed* into thirty-five equal-sized parts and I am considering four of those equal-sized parts at this time. The fraction of her life that she spent in high school is four-thirty-fifths. The fraction of her life that she spent outside of high school is thirty-one-thirty-fifths.

MEDIATOR: Very good! Now, explore this page.

Both fractions must be considered to be produced from the same whole to provide a *logical basis* through which we can *form a relationship* between the two fractions. The whole can be *analyzed* into equal-sized parts by *analyzing* each part of a fraction into an integral number of equal-sized parts. The least common multiple (LCM) is the smallest equivalent result of *analyzing* the whole this way from the two fractions we need to consider for the relationship. *Analyzing* each one-half of the whole into three equal-sized parts is equivalent to *analyzing* each one-third of the whole into two equal-sized parts. This *provides the logical evidence* we need to *form a meaningful relationship* between the two fractions.

MEDIATOR: Let's *analyze* this page (see Figure 5.8).

STUDENT 4: There are circles at the top and bottom of the page. They appear to be the same size. Because one-half is at he top and one-third is at the bottom, the circle must represent the common whole that one-half and one-third are produced from.

STUDENT 11: At the top I see: "Number of equal-sized parts the whole has been *analyzed* into." Under the first circle I see the number two. This must mean that the common whole was *analyzed* into two equal-sized parts.

STUDENT 9: In the next rectangle I see "the number of equal-sized parts each half or each third is *analyzed* into." Under the number two for one-half I see one. I think that this means that each half of the whole is already one part and there are two parts for the whole. For one-third I see three, which means that each one-third of the whole is already in one part.

STUDENT 6: When we move across to the next position I see that when each one-half is *analyzed* into two equal-sized parts, the total number of equal-sized parts in the whole becomes four. Now when each one-third is *analyzed* into two equal-sized parts the total number of equal-sized parts in the whole becomes six.

STUDENT 12: This is really interesting. I see that these two analyses are working together between the two fractions.

$^1/_2$

Number of equal-sized parts the whole has been analyzed into:

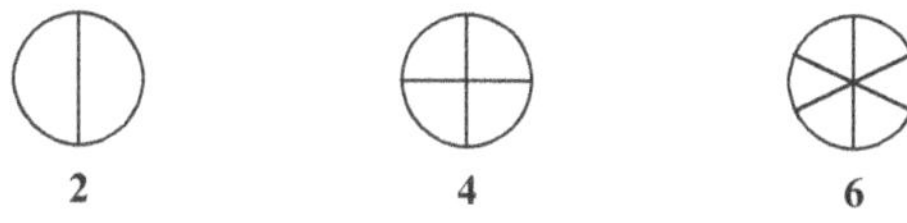

2 4 6

Number of equal-sized parts each $^1/_2$ and each 1/3 is analyzed into:

1 2 3

1/3

Number of equal-sized parts the whole has been analyzed into:

3 6

Figure 5.8. Part and whole.

STUDENT 16: When I *analyze* each half into three equal-sized parts the total number of equal-sized parts in the whole becomes six. This result is the same number of equal-sized parts that were obtained when each third was *analyzed* into two equal-sized parts.

MEDIATOR: This is very powerful! I want to truly thank each one of you for *being very precise* in your thinking and in your language.

STUDENT 13: Using precise language helped me to organize my thoughts and think more clearly. It has been a big struggle but it has been worth it.

MEDIATOR: Is this part of the rigor that you are expressing?

STUDENT 10: Yes, indeed. This is the challenge that I have to dig in and meet.

STUDENT 5: I think that the rigorous thinking is helping me to improve my language. As I gain understanding I am building vocabulary to help me to describe this understanding in math and in life in general.

MEDIATOR: Am I hearing you correctly? Are you saying that your use of precise language is helping you to think rigorously and that your rigorous mathematical thinking is helping you to develop more precise language?

STUDENTS: Yes!!

MEDIATOR: Where do we go from here?

STUDENT 14: What I have gotten out of this so far is that *analyzing* each half of the whole into three equal-sized parts gives the same result as *analyzing* each third of the same whole into two equal-sized parts. The total number of equal-sized parts in the whole will be six in each case. The fact that this is the same number of parts in each case makes this result the least common multiple.

MEDIATOR: Good summary but not completely correct. The fact that it is both the same number and the smallest number possible for this process makes it the LCM.

STUDENT 14: Yes. I see.

MEDIATOR: Let us examine what we learned from finding the LCM. You said that *analyzing* each half of the whole into three equal-sized parts gives the same result as *analyzing* each one-third of the same whole into two equal-sized parts. Let's go further. How does it benefit us to know this?

STUDENT 2: This *provides me with a relevant cue.* Manuel said that the analysis of the whole between the two fractions is working together.

MEDIATOR: What do you mean by "working together"?

STUDENT 2: The least common multiple is six. Because this means that the whole is *analyzed* into six equal-sized parts we have to move to looking at this whole being *analyzed* into six equal-sized parts and at the same time seeing the whole as two equal-sized parts and three equal-sized parts.

STUDENT 16: It's like translating something from your language to a new language. You have to use both languages.

STUDENT 1: But with the fractions we have three languages. We have to know how to communicate in all three languages.

MEDIATOR: Great thinking!! Tell us, what are the three languages?

STUDENT 1: Halves, thirds, and sixths.

STUDENT 5: We have to translate from halves to sixths and from thirds to sixths at the same time.

STUDENT 9: In translating from halves to sixths, knowing that each half is *analyzed* into three equal-sized parts helps me to know that for each half there are three of the needed parts for this translation.

STUDENT 7: This suggests that by multiplying each part of the fraction by three I will get the value of one-half because the whole was *analyzed* into six equal-sized parts. This helps me to translate halves into sixths.

MEDIATOR: Who will go to the board and write this out?

STUDENT 7: I will. [Student 7 goes to the board and writes the following: $1/2 = 1/2 \times 3/3 = 3/6$.]

MEDIATOR: Explain this.

STUDENT 7: A denominator of two tells us that the whole has been *analyzed* into two equal-sized parts. I need to multiply the denominator by three because there are three new parts in each of these halves. The numerator tells us that we are considering one of these halves at this time. I need to multiply the numerator by three also, because there are three new parts for each half.

STUDENT 3: I see something. The new parts are sixths. Each half is three-sixths.

MEDIATOR: Good. What about the translation for one-third?

STUDENT 6: Let me put it on the board. [Student 6 goes to the board and writes the following: $1/3 = 1/3 \times 2/2 = 2/6$.]

STUDENT 6: Each third contains two new parts. The denominator three in one-third means that the whole has been *analyzed* into three equal-sized parts. I need to multiply this denominator by two and the numerator by two to make the translation. Therefore, three of these thirds contain six of these new parts altogether. Three thirds is the same as six sixths.

STUDENT 8: And one-third translates into two-sixths.

MEDIATOR: Very good!! This has been a powerful discussion. Who can summarize all of this for us?

STUDENT 11: To add or subtract fractions with unlike denominators, we have to *form a logical basis* for *forming a relationship* between the fractions. This means that, first, we have to see that all of the fractions are coming from the same whole. Next we have to find out how many parts each of the equal-sized parts from each fraction will be *analyzed* into so that doing this for each fraction gives the same number of parts in the whole. The number of parts in the whole will be the LCM. We can now use this information to translate each fraction into the fraction with the least common denominator.

MEDIATOR: This is excellent. I'm going to pass out group assignments on adding and subtracting fractions. For each item show all of the details of your work and explain how you used your cognitive functions to do it.

The learner understands that the whole is analyzed into three equal-sized parts for one-third and must be analyzed again into two equal-sized parts for one-half. The learner also understands that the least common multiple (LCM) tells us the number of equal-sized parts the whole must be reanalyzed into to provide the logical evidence to form a transitive relationship between the two fractions.

The learner now understands that the common language of the whole (CLW) can be expressed as $2/2 = 3/3 = 6/6$ and the common language of the parts of the whole (CLP) can be expressed in the forms of the following statements:

If $1/3 \times B = ?/6$ and
$\frac{1}{2} \times G = ?/6$
and $B = 2/2$ and $G = 3/3$;
then $1/3 = 2/6$ and $\frac{1}{2} = 3/6$;
therefore $2/6 + 3/6 = 5/6$

The set of equations above engages the learner in both mathematical syllogistic thinking and mathematical transitive relational thinking.

Conceptual Formation in Algebra Through RMT

When students are asked, "What is the difference between basic mathematics and algebra?" a typical response is that algebra is more difficult than basic mathematics. To the question "What makes algebra more difficult?" one answer that is always given is "Algebra deals with variables and this is more difficult while basic mathematics deals only with numbers." When asked, "What is a variable?" most learners say that a variable is a letter that takes the place of an unknown number. Although the latter statement is partially true, it misses the dynamic nature of a variable and the critical attributes that make algebra and calculus so exciting.

RMT teaching of algebra is designed to engage students in consciously and deliberately practicing the formation of conceptual elements of a mathematical function by the joint use of psychological tools and cognitive functions. These conceptual elements that are grounded in the dynamic nature of variables are (1) change within the context of conserving constancy; (2) changeability; (3) interdependence; (4) cause/effect relationship; (5) input/output relationship; (6) functional relationship; (7) independent and dependent variables; (8) one-to-one correspondence; (9) ordered pairs; (10) slope and intercepts; (11) change in slope; and (12) representation of functional relationship through formulae, tables, and graphs. Not only is the conceptual nature of algebra anchored in the dynamic feature of a variable but its intrinsic beauty stems from this aspect as well. Without grasping the dynamic aspect of a variable algebra becomes a mechanical manipulation of meaningless symbols and dull routines.

Students are guided to start constructing the dynamic concept of a variable through a series of tasks similar to what is given below.

Situation I: $a + 5 = 8$; Situation II: $a + 12 = 19$

1. *Analyze* and *compare* these two situations.
2. Is there a variable in each situation? ________ If so, name the variable. _____

What happened to the variable a as we moved from situation I to situation II? __________________________

A variable is something that ________ its ________ or its ________ or its ____________________ in ________ ____________ or at ________ ________.

Here students develop the understanding that a variable is something that changes its amount or its quantity or its value at different times or in different situations.

Next, learners are given a page that contains a frame with two or three geometric figures serving as models followed by frames of clouds of unconnected dots with various amounts of spacing among them (see Figure 5.9). Each frame contains the number of dots needed to make the models. Learners are mediated to complete the items below:

1. What is *conserving constancy* about the dots only as we move from frame to frame? ____________________.
2. What is changing about the dots only as we move down from frame to frame? ____________________. Therefore, ________ is a ________ because it is ______________________________.

These two questions are designed to help the learner experience what it means to conserve constancy in the context of change. It is important to

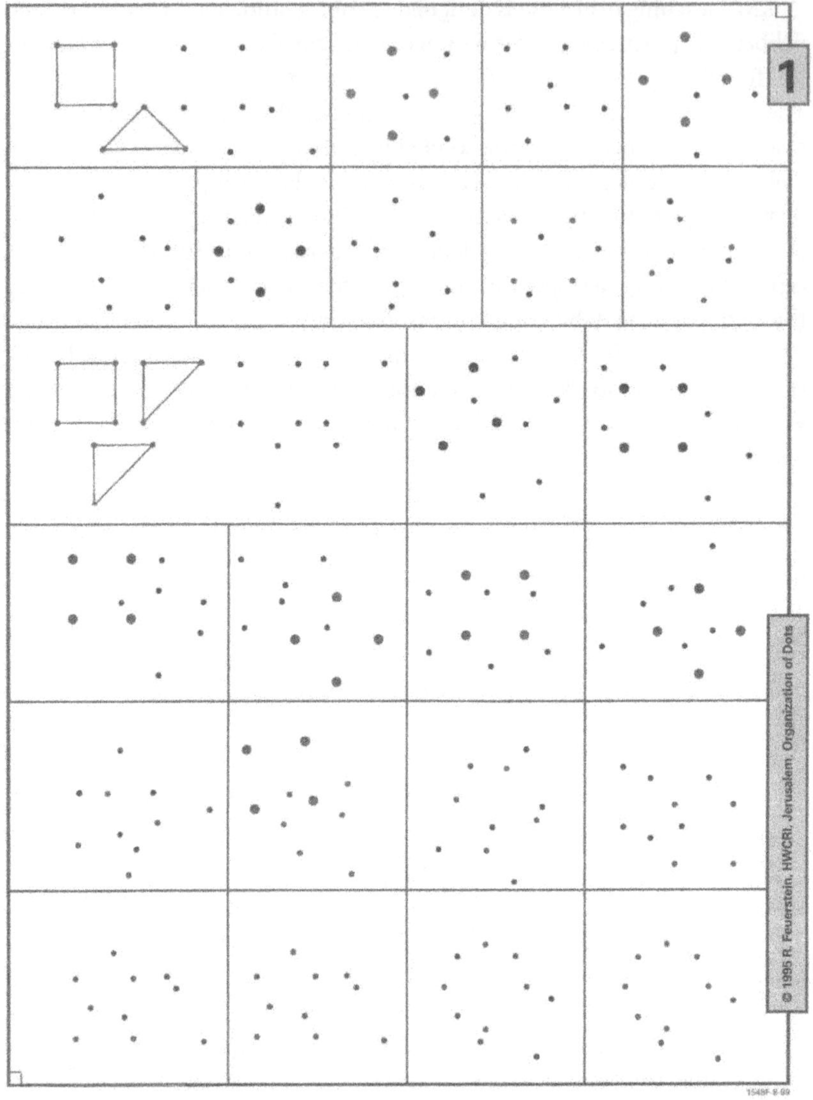

Figure 5.9. Instrumental Enrichment Program, "Organization of Dots." © R. Feuerstein, HWCRI, 1995. Used with permission.

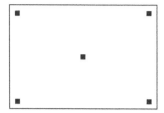

Figure 5.10. Diagram of lowest proximity.

see that the quantity of dots stays the same while the closeness of the dots changes. The closeness or proximity of the dots versus the space between the dots makes the concept somewhat abstract to perceive. Mediation of this concept may be achieved by having five students stand in such a way as to show the lowest possible proximity of students standing on the rectangular floor in the classroom. Students who are sitting are asked to coach the five standing students to arrange themselves in the appropriate positions. After coaching and mediation, the next step is to have one student stand in each corner of the room while the fifth student stands in the center (see Figure 5.10). The teacher tells the students to encode the quantity of this proximity as G and asks the students what value they would assign this proximity because it is the lowest possible. After some intense discussion the students conclude that the value should be zero. When asked, "Why zero?" some students respond that zero is the lowest value possible without one or more of the students going outside of the classroom. If one or more of the students go outside the classroom the value of the proximity would be negative. This opens up a rich discussion regarding why there are rules in mathematics.

The teacher then instructs the students to increase the proximity by 25%, 50%, and 90% and encode the situations as A, C, and B, respectively. They then are asked to show a size relationship of proximity for the four situations using the sign > or <. After some struggle students write the proximity of standing students in G < the proximity of standing students in A < the proximity of standing students in C < the proximity of standing students in B and the proximity of standing students in B > the proximity of standing students in C > the proximity of standing students in A > the proximity of standing students in G. Thus, students conclude that the quantity of the proximity is the opposite of the space between the students or the dots. In addition, they conclude that the proximity of the dots is a variable because it changes its value at different times or in different situations.

Students are now required to connect the dots to make the same size and shape as the models (see Figure 5.9). They are instructed that all the dots should be used and that the same dot cannot be shared by two or more figures.

3. What is *conserving* constancy about the figures as we move from frame to frame?____________________.
4. What about the figures is changing? _______. Therefore, the is a ________ because the__________________________________.

Conserving constancy is necessary to detect change, a critical attribute of a variable. This becomes the first issue to be examined by the student when presented with a mathematical function. When the meaning of *conserving constancy* is determined the nature of the variation is more sharply defined. Understanding that the shape and size of the figures are *conserving constancy* allows the learner to observe that the overlapping of the figures changes as we move from frame to frame (Figure 5.11).

Once the learner *provides logical evidence* that the proximity of the dots and the overlapping of the figures are variables, the question can be asked regarding whether there is a relationship between these two variables. The answer to this question can be determined by examining the various frames sequenced from the lowest proximity of the dots to the highest proximity of the dots in the clouds of unconnected dots and the sequenced frames from the lowest overlapping of the figures to the highest overlapping of the figures in clouds of dots that are connected. The learner now is asked to state this relationship in words orally and in writing. It can be observed that as the proximity of the dots increases in its value or its quantity or its amount the overlapping of the figures increases in its value or its quantity or its amount, which is a verbal formation of the functional relationship between the two variables (see cognitive function *forming a functional relationship* in Table 4.1).

The question is now raised does one variable depend on the other, and, if so, which variable depends on the other? Because the proximity of the dots existed before the overlapping of the figures, the overlapping of the figures depends on the proximity of the dots. Learners are confronted with the following three questions: Is this an independent/dependent relationship? Is this a cause/effect relationship? Is this an input/output relationship? Now they are required to label the variables based on these three types of relationships as the independent variable being the cause variable or the input variable and the dependent variable as being the effect variable or the output variable. Using more symbolic language with the overlapping of the figures encoded as OF and proximity of the dots encoded as PD the *forming a functional relationship* between the two variables is expressed as OF = *f*(PD). Using more general mathematical language with PD being x and OF

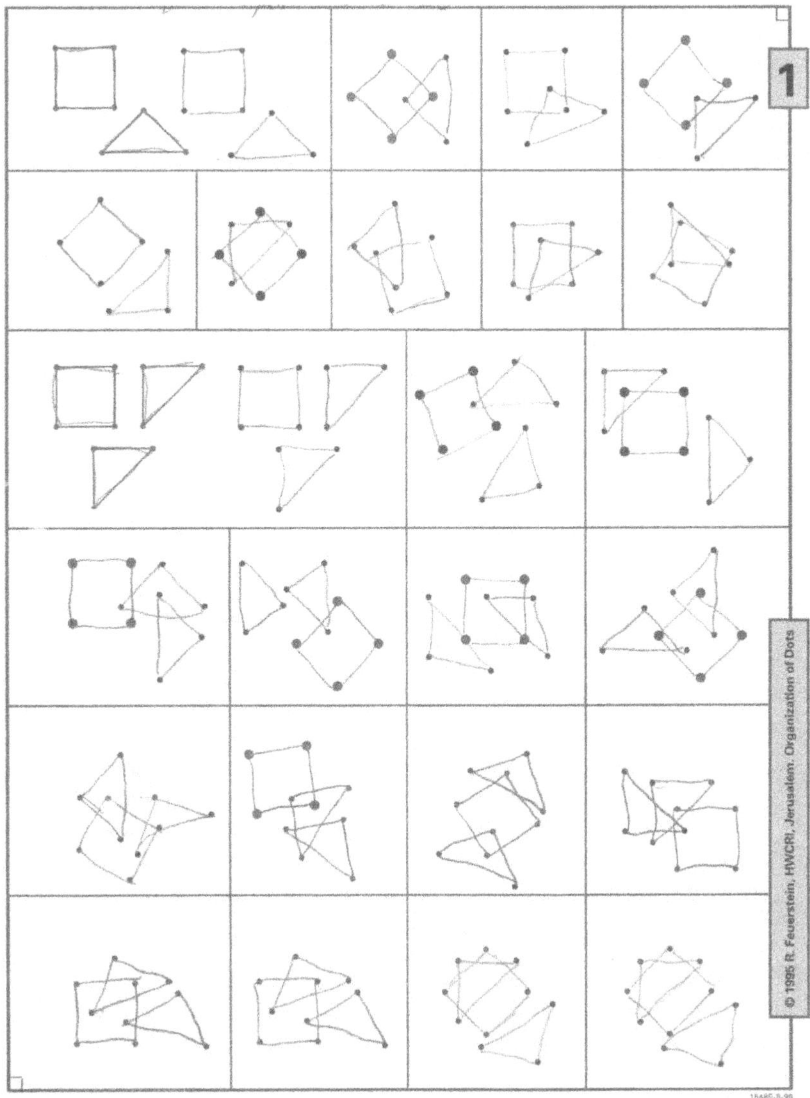

Figure 5.11. Student's solution of "Organization of Dots" tasks.

being y the *functional relationship* between the two variables is expressed as $y = f(x)$.

The class is provided further practice in forming these conceptual understandings by mediation through tasks shown in Figures 5.12 and 5.13. The independent variable or cause variable or input variable in the first case (see

Figure 5.12. Functional relationship between two variables.

Figure 5.12) may be the amount of time the candle burns or the amount of heat produced from the flame while the dependent variable or the effect variable or the output variable may be the length of the candle or the amount of wax in the pan of the candle holder. In the second case (see Figure 5.13) when moving from left to right the independent variable or cause variable or input variable is the size of the square, whereas the dependent variable or the effect variable or the output variable is the content of the square. Mediation through these tasks provides learners with experiences to move from pictorial and figural presentation of events to theoretical derivation of symbolic expressions of functional relationship between two variables.

The students are then presented with a mathematical equation containing two variables, such as $y = 3x + 4$, and the first consideration is what is *conserving constancy*. The superficial but often given answer is that numerals 3 and 4 are *conserving constancy*. However, a deeper examination will lead students to state that the functional relationship between the two variables conserves constancy. They are mediated to express this functional relationship verbally using mathematical language as follows: "For every value of the independent variable x we multiply it by 3 and to that product we add the value 4 to get the corresponding value of the dependent variable y."

At this point students are taught how to appropriate a table as a mathematically specific psychological tool that helps to organize and form relationships

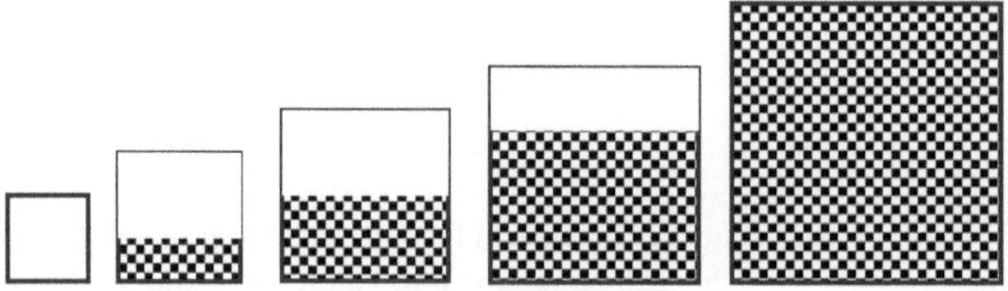

Figure 5.13. Functional relationship between size of square and content of square.

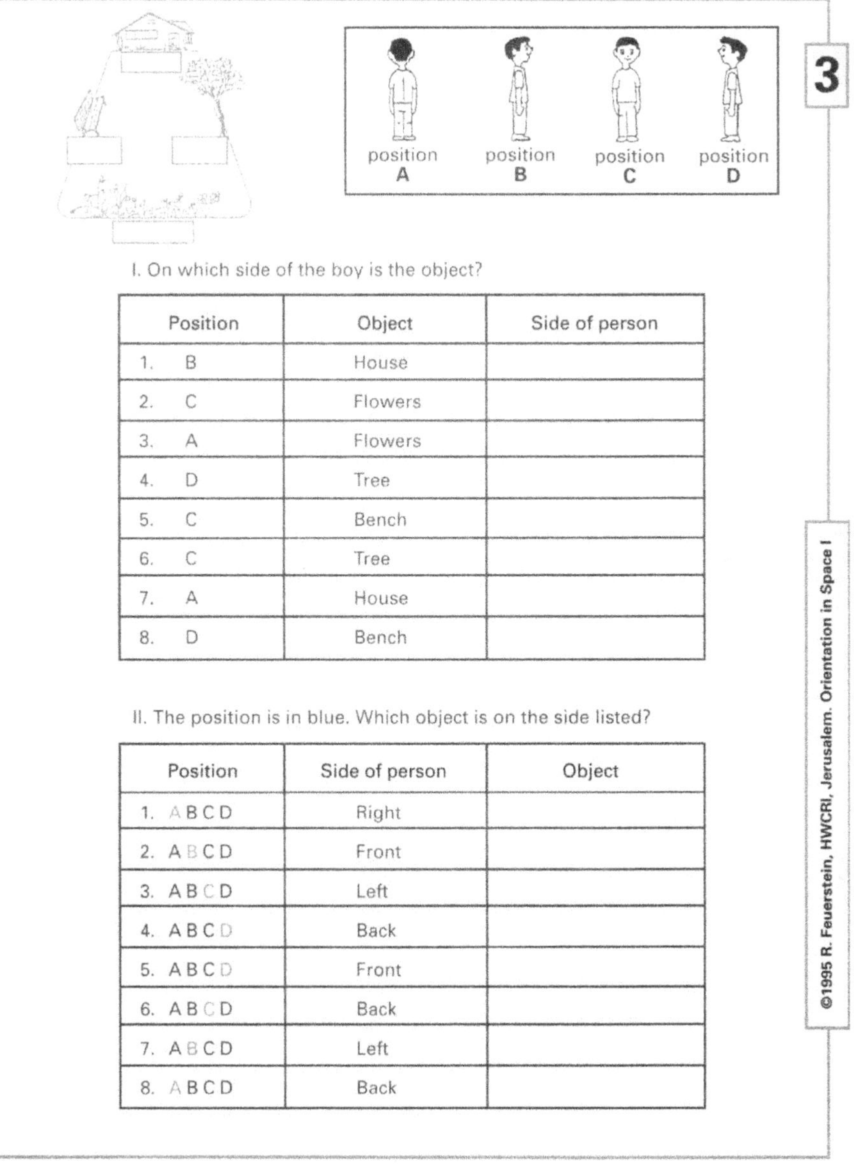

3

position A | position B | position C | position D

I. On which side of the boy is the object?

Position	Object	Side of person
1. B	House	
2. C	Flowers	
3. A	Flowers	
4. D	Tree	
5. C	Bench	
6. C	Tree	
7. A	House	
8. D	Bench	

II. The position is in blue. Which object is on the side listed?

Position	Side of person	Object
1. A B C D	Right	
2. A B C D	Front	
3. A B C D	Left	
4. A B C D	Back	
5. A B C D	Front	
6. A B C D	Back	
7. A B C D	Left	
8. A B C D	Back	

Figure 5.14. Instrumental Enrichment Program, "Orientation in Space." © R. Feuerstein, HWCRI, 1995. Used with permission.

between values for the variables. The following tasks, along with Figure 5.14, are used to bring about this appropriation.

1. What is the name of the two things that exist from one fourth of the page (see Figure 5.14) to the bottom? ________. What gives something structure?

____

Let us examine the first table on this page. The words, letters, and numbers inside the table are the content of the table. Ignore all of the content at this time. What gives this table structure? ________________________. When we move in a straight up or down direction in space, what do we call this orientation? ________. What do you call the spaces that run ________________ in a table? ______________ When we move straight from the left to the right direction in space or from the right to the left direction in space, what do we call this orientation? ________ What do we call the spaces that run ________ in a table? ________

2. The structure of every table is made up of ________ and ________. The ________ run ________ and the run ________. Each ________ has a title or a ________ or a ________ that ________ or ________ the things under it into a ________.

If you were told to put the items apple, orange, pineapple, grape, banana, and pear in the same column, how would you label the column?________ Why?

____

________________________What is more general, the heading of the column or the items under the heading?____________________. So what is the heading doing to the items under the heading?

____

________.When you were using your mind to develop the label for the column with the items apple, orange, pineapple, grape, banana, and pear, were you going from the general to the specifics or were you going from the specifics to the general? ________

You were just using a new cognitive function. Let us write it in our RMT journal. The name of this cognitive function is *inductive-deductive thinking.* What action are you taking in your mind when you are doing *inductive thinking*?

____

You also were taking another thinking action in your brain when you gave this group of items a more general name. That cognitive function was *generalizing.* Write this new cognitive function in your journal.

Let's now define it by describing what we are doing when we take this thinking action: placing two or more items into a _______ based on the logical evidence that they share certain ______________________________.

3. How many columns are in the first table on this page? Describe what each column is doing.

What cognitive functions are you using and describe how you are using them?

4. How many different positions are there? _______ What are the various positions? _______. The position of the boy is changing its quantity at different _______. This _______ that the position of the boy is a _______. How many specific objects are there in this table? _______ What are the specific objects? ______________. The object is changing its _______ at different __________. This __________________________ that the object is a _______.

5. Let's go to the 1 in the first row of the first table. As we move from left to right, we take the value "B" of the variable position and look at the value "house" for the ____________ object and ______________________________ between "B" and "house" and see that the _____ of the boy. In the third _______ of the table we are ____________________ between the "a" value for the variable _______________________________ and a specific value for the variable _____________. Once we choose the value "B" for the variable position and the value "house" for the variable object, the _______ between _______ and _______ becomes fixed or _______. We now have a pair of values, one value for the _______ ____________________ and another value for the _______ _______. The value for the _______ is _______. The value for the _______ is _______. The value _______ corresponds with the value _______. What does this mean?

Because the value ____ for the _______________ corresponds with the value _______ for the _______ B, house is an _______. An _______ exists when two _______ values from two variables _______ with each other.

6. Every tool has a structure and a function or a _______ or a _____________.

Does the structure of the tool bring about its function or does the function of the tool bring about its structure? ___________________________________ _____________. Therefore, every tool has a _______________________________ _____________________. Is a table a tool? If so, what is your logical evidence?

Table 5.7. Use of a table as a mathematically specific psychological tool to organize and form relationships between data $y = 3x + 4$

Independent variable x	Dependent variable y	Ordered pair x, y	Change in x, Δx	Change in y, Δy	Unit functional relationship slope $\Delta y / \Delta x$
0	4	0,4			
			1	3	3
1	7	1,7			
			1	3	3
2	10	2,10			
			2	6	3
4	16	4,16			
			3	9	3
7	25	7,25			
			5	15	3
12	40	12,40			

Learners are now mediated to construct a table (see Table 5.7) to organize and form relationships between the values for the variables in the equation $y = 3x + 4$.

Students are required to verbalize the functional relationship between the two variables: For every value of the independent variable or the cause variable or the input variable x we multiply it by 3 and to the product we add the value 4 to get the corresponding value of the dependent variable or the effect variable or the output variable y. By utilizing this verbalization of the functional relationship between the variables the data in the first two columns can be generated. The data in the third column, the ordered pairs, are produced by forming relationships between each value for x and the corresponding value for y. The set of ordered pairs *forms the functional relationship* between the two variables for $y = 3x + 4$. The data in the fourth and fifth columns show the changes in the values for x and y, respectively. The data in the sixth column *form the unit functional relationship* that *conserves constancy* and *provides mathematical logical evidence* that this is a linear function. This unit functional relationship, which is the same as the slope, comes from making a connection between the change in the amount of the dependent variable that is produced by a unit change in the amount for the independent variable that is defined by the functional relationship between the two variables expressed in the mathematical function or the algebraic equation.

At this point students are mediated toward appropriation of the x-y coordinate plane as a mathematically specific psychological tool using the tasks below.

What does the word *plane* mean in mathematics? ____________________

____________________________. How many dimensions exist in a plane? __________ How many dimensions exist in a straight line? ______ How many dimensions exist in a cube or a sphere? ____________.

Draw a straight vertical line segment of about 6 inches in the middle of the space below. Be as precise as you can without using a ruler.

Make a number line out of the horizontal line by analyzing it into 12 equal size parts. Encode the origin in the middle. Encode the first point to the right of the origin with the number 2. Encode the other points on the number line.

What a number line does as a tool is called its ______ or its______ or its ______. In algebra, one ______ of the number line is to ______ and ______ the ______ or the ______ or the ______ of a ______ and ________________ between these __________ or ______ or __________. What variable will be encoded on this number line? ______

What kind of variable is this? ___________________________. Construct a vertical number line that intersects this number line at the origin. Encode the first point above the origin with the number 4. Encode the other points. What variable will be encoded on this number line?______ What kind of variable is this? _________________. The ______ or ______ or ______ of this ______ is to ______ and ______ the ______ or the ______ or the______ of the______ ______ and______ between these ______ or ______ or ______.

What does the syllable "co" mean in the word "coordinate"? ______ or ______. When two things are coordinated this means that they have been arranged in the same order according to rank. Suppose 10 students in group Z form a straight line on one side of the classroom and start counting off from one end: one, two, three, four, and so on. Now suppose a different group of 10 students, group K, form a different line somewhere else in the classroom and start counting off from one end: A, B, C, D, and so on. Although the two lines of students may not be right beside each other, student A in group K has the same rank and order of importance as student 1 in group Z. Student 3 will have the same rank and order of importance in group Z as student ______ in group K. Student J in group K will have the same order and rank of importance as student ______ in group Z, and so on. Therefore, coordination in mathematics means ____________________

____.

One number line organizes the _______ or _______ or _______ for one _______. The other number line organizes the _______ or _______ or _______ for the other _______.

The *x-y* coordinate plane can now be used as a mathematical psychological tool to _______ the _______ of the two _______.

When a value for the _______ corresponds with the _______ for the _______ we write the value for the _______ first and then we write the value for the _______ and separate them with a comma. This group is an _______. When we do this we are using the cognitive function _______ because we are _______.

The *x-y* coordinate plane has four _______. In the first _______ we coordinate only _______ for the _______ *x* and _______ for the _______ *y*. In the second _______ we coordinate only _______ for the _______ *x* and _______ for the _______ *y*. In the third _______ we coordinate only _______ for the _______ *x* and _______ for the _______ *y*. In the fourth _______ we coordinate only _______ for the _______ *x* and _______ for the _______ *y*.

One purpose for mediating students through these tasks is to help them understand that the *x-y* coordinate plane is a two-dimensional surface that is structured by two number lines intersecting to form four right angles that can be used to organize and form relationships between the values of two variables. The purpose or use or function of each number line is to organize and compare the values or quantities or amounts of a variable and form relationships between these quantities. Thus, the two number lines together can be used to coordinate values of the independent variable with values of the dependent variable. Each pair of coordinated values for the two variables will be the ordered pair, which means that one value for the independent variable will have a corresponding value for the dependent variable.

Now that the *x-y* coordinate system has been constructed learners are required to use it as a mathematical tool to reorganize the data formed in the table for $y = 3x + 4$ and reconstruct the functional relationship between the variables *x* and *y*. The results of one student are shown in Figure 5.15.

Next, students are given the following tasks.

Compare the equation, the table, and the graph. What is conserving constancy in the (a) equation? ____

____

(b) table? ____

_______________________________ (c) graph? _______________________________

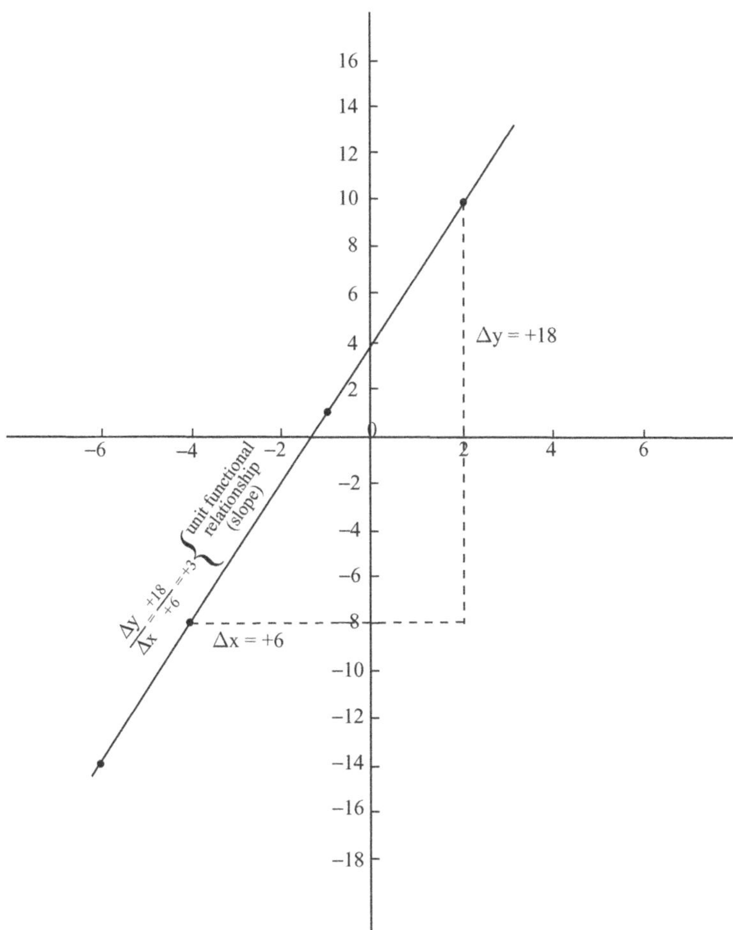

Independent Variable x	Dependent Variable y	Ordered Pair (x, y)	Change in y Δy	Change in x Δx	Unit Functional Relationship $\Delta y/\Delta x$ (Slope)
−6	−14	(−6, −14)	-	-	-
−4	−8	(−4, −8)	+6	+2	+3
−1	+1	(−1, +1)	+9	+3	+3
+2	+10	(+2, +10)	+9	+3	+3
+5	+19	(+5, +19)	+9	+3	+3

Figure 5.15. Graph and quantitative data for the functional relationship $y = 3x + 4$.

What is changing in the (a) equation?

(b) table?

(c) graph?

Describe the order-pairs and tell how they appear in the (a) equation.

(b) table.

(c) graph.

Describe the functional relationship in the (a) equation.

(b) table.

(c) graph.

Provide logical evidence for the advantages and disadvantages of the following three mathematical psychological tools for expressing the functional relationship between the two variables: (a) $y = 3x + 4$; (b) the table; (c) the graph.

6 RMT Application, Assessment, and Evaluation

The application format for mathematics concept formation through Rigorous Mathematical Thinking (RMT) involves three factors – topic, grade level, and time of application. Although the RMT format consists of different topics or mathematical concepts, each requires learning that involves six core mathematical concepts – quantity, relationship, representation, abstraction/generalization, precision, and logic/proof. Each topic also involves the appropriation and use of the mathematically specific psychological tools of signs and symbols and mathematical language. Because there are cognitive functions that are naturally aligned with and needed to build the core concepts, teaching the topics will require development of these cognitive processes. Teaching a certain topic provides a supportive foundation to facilitate teaching the topic that follows. However, because each topic demands its own core conceptual development along with the naturally aligned cognitive functions, the topics can be taught relatively independently of each other.

Examples of the amount of time required for RMT classroom teaching to develop and/or improve understanding and skills with regard to specific concepts/topics at respective grade levels are given below.

- Number sense for 2nd to 7th grades, 30 hours.
- Basic math operations and properties for 2nd to 7th grades, 40 hours.
- Meaning of fractions, adding, subtracting, multiplying, dividing fractions with like and unlike denominators, simplifying fractions, converting from fractions to decimals and vice versa, and ratio and proportions for 4th to 7th grades, 45 hours.
- Factoring, writing expressions, and solving simple equations for 4th to 7th grades, 20 hours.

- Basic geometric properties of two-dimensional and three-dimensional figures, area, and volume for 4th to 7th grades, 25 hours.
- Prealgebra concepts for 4th to 7th grades, 40 hours.
- Linear equations, inequalities, and linear functions for grade 8 through high school, 45 hours.
- Nonlinear equations, inequalities, and nonlinear functions for grade 8 through high school, 60 hours.
- Geometry that includes points, lines, planes, angles, developing proofs, area, volume, and so on for high school, 95 hours.

The optimum length of each classroom session and frequency for producing deep understanding and efficient mathematical skills are 90 minutes with three sessions per week. The minimum session length and frequency are 45 minutes with three sessions per week.

Lesson Planning

Because RMT focuses on teaching and learning big ideas the lesson planning process is very important to success of program implementation. This is true both for each individual lesson as well as for a cluster of lessons that may be considered as a conceptual unit. RMT lesson planning presupposes that teachers examine their educational practice beliefs as well as their proficiency in mathematics, including its language and conceptual organization. Planning an RMT lesson can be envisaged as an ongoing metacognitive process characterized and articulated through series of questions that initiate and sustain higher order thinking first in the teacher and then in the students.

The RMT teacher is first and foremost a mediator. (See Chapter 4 for the description of the mediated learning.) The lesson planning requires the teacher-mediator to engage in rigorous metacognition; the RMT instruction must be strategic. The teacher must understand and be able to visualize the dynamics of the interactions among the student, the learning material, and its context and the teacher. This understanding should be transformative in nature and form the framework for lesson planning. Rigorous metacognitive thought primarily takes the form of the teacher asking him- or herself a number of basic questions:

- What do I know about my class as a whole and about individual students?
- What is my intention and how best do I engage each student so that they share my intention?

- What cognitive functions align with what core mathematical concepts? How is each concept related to what big idea, and how are both related to what standard?
- What are the most important skills and knowledge I need to teach this concept to my group of diverse learners?
- How do I measure cognitive growth in my students?
- What do I do if a student or a group of students are having difficulty understanding any part of the process?
- What are the major mediated learning criteria needed to guide this unit or lesson?
- How will I grade the tasks or activities in the unit or lesson?

The structure and components of the RMT process and the process itself provide valuable elements for developing unit/lesson plans. RMT unit/lesson planning requires coherency between the phases and steps of the process. To achieve a "flow within the process," teachers must be able to analyze and integrate the cognitive processing, psychological tool building, and content (including applications) components. In addition, the assessment/evaluation of the learning experience must be developed and articulated in the lesson plan before any portion of the plan is implemented.

Conducting a structural analysis of the content to be learned and the RMT process is a necessary step to examine relationships between the components of the process and mathematics content – cognitive functions to psychological tools to basic mathematics concepts to mathematically specific psychological tools for cognitive conceptual construction to applications of cognitive conceptual construction. Constructing an operational analysis of the student/teacher interaction in the context of the learning process and materials facilitates teacher development of questions that guide instruction (mediation), assessment and evaluation, and teacher reflection of the lesson planning process.

Finally, teacher collaboration in the planning process is preferable to planning alone. Collaboration facilitates discussion, reflection, and critique of the process and in itself is educational for each participant to improve teaching. In their recent article on mathematics teaching O'Donnell and Taylor (2006/2007) quoted Shulman's (1986, p. 13) idea that "teaching is a process of 'pedagogical reasoning and action' that involves the need for teachers to grasp, probe, and comprehend an idea, to 'turn it about in his or her mind, seeing many sides of it. Then the idea is shaped or tailored until it can in turn be grasped by the students'. . . . Teachers also need to develop strategic knowledge to confront troublesome, ambiguous teaching situations and build a 'wisdom of practice.'"

Mediation During the Lesson

The key to successful implementation of the RMT model is the quality of the teacher's mediation. The following vignette demonstrates how Cherise Copeland, a 3rd-grade teacher, introduces her students to cognitive functions simultaneously with working on the mathematics topic of rounding to the nearest 10. During the previous lessons with Instrumental Enrichment (IE) tasks the students have already been familiarized with such cognitive functions as activating prior knowledge, comparing, inferential-hypothetical thinking, and providing logical evidence.

TEACHER: [holding up a pencil] What is this?

STUDENT: A pencil.

TEACHER: What is this used for?

STUDENT: We use it to write.

TEACHER: How is this pencil made?

STUDENT: There is lead inside the wood with an eraser on the back.

TEACHER: What word can we use describe how something is made?

STUDENT: The way it's made.

TEACHER: How about the way a building is made?

STUDENT: It's made with wood and brick.

TEACHER: What do you call this material all put together?

STUDENT: It's structure.

TEACHER: This pencil is made a certain way. [holding up an eraser] Can I use this to write?

STUDENT: No, it has no lead in it. You need the lead to write.

TEACHER: Very good. A pencil has a specific function. What other word could we use for function?

STUDENT: The pencil has a certain use.

TEACHER: From looking at this pencil, we can determine that its structure, the way it is made, is related to its function, what it is used for. Can you give another example of something we use in math that has structure/function relationship?

STUDENT: A ruler. It is a straight piece of plastic with numbers on it. We use the numbers on it to measure how long something is.

TEACHER: [Draw a number line on the board, without the numbers.] Describe the structure of this.

STUDENT: It is a long straight line, with a lot of smaller lines.

TEACHER: What can you tell me about the little lines?

STUDENT: They are the same distance apart.

TEACHER: What can we use this structure for?

STUDENT: To make a number line.

TEACHER: How did you know that?

STUDENT: I have used one before.

TEACHER: What cognitive function did you use to know this was a number line?

STUDENT: Activating prior knowledge, I remember using one before.

TEACHER: When have you used a number line?

STUDENT: When counting, adding, or subtracting.

TEACHER: Is this structure important to its use or function? Why?

STUDENT: Yes, because the numbers stay in order and it makes it easier to compare the quantities.

TEACHER: How do I place the numbers on the number line?

STUDENT: In order from smallest to biggest.

TEACHER: I am going to give each little line a name, starting with 0 and ending with 10. What is that called when you give something a name based on its critical attributes?

STUDENT: Labeling.

TEACHER: What is it called when we are giving something a code, when you are putting meaning into a code?

STUDENT: Encoding.

TEACHER: Why do we need to encode the little lines?

STUDENT: So we know what numbers to use.

TEACHER: Now that I have my number line encoded, let's use it. If I have eight dollars, ABOUT how much money do I have?

STUDENT: Ten dollars.

TEACHER: How did you know that?

STUDENT: Because the eight is closest to the ten.

TEACHER: If the eight is closer to the ten then the answer must be ten. What cognitive function are we using here?

STUDENT: Inferential-hypothetical thinking.

TEACHER: How do you know the eight is closer to the ten?

STUDENT: There are less numbers between the eight and ten than there are between the zero and eight.

TEACHER: What cognitive function are you using when you are looking at two different things?

STUDENT: Comparing.

TEACHER: What did you compare?

STUDENT: How many numbers are between eight and ten to how many numbers are between zero and eight.

TEACHER: What is that called when we try to figure out ABOUT how much something it?

STUDENT: Estimation.

TEACHER: Good, in math when we estimate we are also rounding.

[Encode the number line from 10 to 20.]

TEACHER: If I have seventeen apples, ABOUT how many apples do I have? And how do you know that?

STUDENT: Twenty, because the seventeen is closer to the twenty.

TEACHER: To explain your answer, what cognitive function did you use?

STUDENT: Providing logical evidence.

TEACHER: What is another name for twenty?

STUDENT: Two tens.

TEACHER: If I ask you to round, what is it that I want you to round to?

STUDENT: The number that it is closest to.

TEACHER: Seventeen is closer to eighteen than twenty. So why did you answer twenty?

STUDENT: Because the number line ends at twenty.

TEACHER: Why do you think the number line ends at twenty?

STUDENT: Because that is when you stopped.

TEACHER: How do you know where to stop?

STUDENT: Maybe you stopped at twenty because it is an easy number.

TEACHER: What makes twenty an easy number?

STUDENT: It just has tens and no ones. It's easier to count.

TEACHER: Because using just tens is easier to count, what did you round to?

STUDENT: The closest ten.

TEACHER: What is another word for closest?

STUDENT: Nearest.

TEACHER: Good, so you rounded to the nearest ten. [without using a number line] What is twelve, rounded to the nearest ten? Before you answer, what are you thinking about?

STUDENT: I am thinking of the number line.

TEACHER: But there is no number line on the board, so where is your number line?

STUDENT: There is a picture in my head.

TEACHER: What cognitive function are you using when you create a picture in you head?

STUDENT: Visualizing.

TEACHER: Does visualizing help you find the answer?

STUDENT: Yes.

TEACHER: What do you call something you use to help you do something? For example, a hammer can help you build a birdhouse. What is the hammer?

STUDENT: A tool. It helps you put together the birdhouse.

TEACHER: Good. Is the number line a tool?

STUDENT: Yes.

TEACHER: Why?

STUDENT: Because it helps you with ordering and comparing quantities.

STUDENT: And it helps you with rounding.

TEACHER: Is the number line a tool that you have to hold in your hand?

STUDENT: No, you can use it in your head.

TEACHER: Because we really understand and know how to use a number line, it becomes a tool. But because we use it in our head, it becomes a psychological tool. We don't have to draw one, we can just visualize it in our head. So what is twelve rounded to the nearest ten?

STUDENT: Ten.

TEACHER: Can you provide me with your logical evidence?

STUDENT: When I think about the number line the twelve is closer to the ten instead of the twenty.

Assessment

The nature of assessment and evaluation in the context of the RMT process is indeed challenging due to the complexity of RMT instruction. However, assessing and evaluating student progress and proficiency in doing and applying RMT is not impossible. In this section, assessment and evaluation for RMT is defined and delineated within the context of classroom teaching and learning. Questions asked of assessment and evaluation are: (1) What needs to be measured? (2) Which type of assessment framework will be used and when are the assessments to be administered? and (3) What will be the instruments and/or format used?

RMT is composed of three phases or components: (1) enhancement of students' cognitive development, (2) acquisition of content as a process, and (3) cognitive conceptual construction practice. Assessment and evaluation of student performance in these components should reflect a high quality of rigor and coherency with the implemented curricular and instructional components. We define mathematical rigor as quality of thought that reveals itself when learners are engaged through a state of vigilance driven by a strong, persistent, and inflexible desire to know and deeply understand. As a result of this vigilance, metacognitive thinking is produced in the learner as a habit of mind, thus creating a disposition in the learner to constantly seek success and completion.

In general, RMT assessment and evaluation should include measuring progress and proficiency in cognitive and mathematical language usage within the appropriate context; conscious and accurate usage of prior knowledge with the introduction of novel conceptual content; strategic thinking and planning for problem solving in various contexts; the learner's disposition toward rigor and challenge; and the learner's level of class participation and engagement on tasks.

Three general constructs for assessments are considered compatible for RMT implementation. They are formative, summative, and evaluative. Formative assessments can be described as assessments that are informal, ongoing, often embedded in instruction, and usually prospective in their orientation, answering the question "How will students' learning be enhanced?" Summative assessments are typically administered after instruction ends (at the end of a unit, term, or school year) and answer the question, "What do the students understand?" Evaluative assessments quantify student progress and proficiency, usually in the form of points and letter grades that are interpreted on a continuum between qualities of passing and failing (Rothman, 2006).

Evaluative assessments in RMT take into consideration the formative assessments and the summative aspects. Evaluating students' work in an RMT class would integrate quantitative and qualitative aspects of student engagement and work at the academic and nonacademic levels (see Marzano, 2000, 2006).

RMT assessment must reflect the coherency among the curriculum (content), instruction, and the climate and culture of the classroom if it is to be valid. Therefore, because RMT is fundamentally cognitive in nature, assessment instruments should measure the cognitive and metacognitive development in students in the context of content learning. Examples of suitable assessment instruments would be rubrics (especially analytic), essays, journal entries, surveys, performance-based tasks, and case study problem-solving approaches and to have students explain, in writing, their approaches to problem solving.

RMT Program Research in Primary School

The first question to be asked when researching the RMT application is whether this instructional paradigm indeed enhances the students' general cognitive performance **together** with their mathematical performance. The following data were collected in the three 4th-grade classrooms of the public school in a medium-sized Midwestern city. A class consisting of low-performing white, African American, and Latino students was taught the math concepts of fractions and function for 60 hours over a period of 6 weeks by a teacher who received RMT training. Supervision was provided by an RMT expert. During the same period two other 4th-grade classes of similar

Table 6.1. Pre- and posttest cognitive OLSAT scores in RMT and comparison 4th-grade classes

	RMT group ($N = 20$)			Comparison group 1 ($N = 13$)		
	Pretest	Posttest	Gain score	Pretest	Posttest	Gain score
Mean	17.90	24.28	6.38*	26.23	26.08	−0.15
SD	7.73	10.27		9.44	10.97	

*$t = 2.27$; $p = 0.025$.

sociocultural and academic status in the same school were taught the same concepts by regular teachers who received no RMT training.

Table 6.1 shows cognitive gains made by students in RMT and in one of the comparison classes over a period of 6 weeks. The cognitive performance was evaluated by Otis-Lennon School Ability Tests (OLSAT; Otis and Lennon, 1996). The cognitive effect size of RMT intervention is 0.7, whereas the cognitive effect size of regular teaching is −0.007. Though the comparison group demonstrated a much higher cognitive level at the pretest there was no sealing effect. At the same time it is possible that significant cognitive gains were made in the RMT group exactly because their starting point was so low. Thus the only claim that can be made on the basis of this data is that RMT group indeed improved their general cognitive performance. The importance of this improvement can be judged only by comparing it to the progress made by the two groups in mathematical achievement.

Table 6.2 shows the results of algebra and fractions concept tests in the RMT and comparison groups. One can see that the pretest performance level was similar in both groups. The gain of the RMT group was significant and the effect size very large (=2.3). The comparison group made no gain. One may thus conclude that the RMT group indeed made significant progress in mathematical conceptual understanding in parallel to their cognitive gains.

Table 6.2. Pre- and posttest mathematical concepts scores in RMT and comparison 4th-grade classes

	RMT group ($N = 19$)			Comparison group 1 ($N = 14$)		
	Pretest	Posttest	Gain score	Pretest	Posttest	Gain score
Mean	2.23	3.54	1.31*	2.19	2.18	−0.01
SD	0.51429	0.6362		0.4393	0.5789	

*$t = 6.98$; $p = 0.005$.

The comparison group made no gains in either general cognitive performance or mathematical concepts. Moreover, two-way ANOVA analysis confirmed the existence of a significant interaction effect in the RMT group between the change in cognitive performance and the change in mathematics concepts performance ($F = 10.121$; $p < 0.005$). The latter fact points to the connection between cognitive and conceptual gains as assumed by the RMT paradigm.

The next and crucial question is whether the improvement of mathematical reasoning observed in the RMT group immediately after intervention had a lasting effect. This question was addressed by comparing the county standards assessment (mathematics benchmarks) results in the RMT and in two comparison groups in the third and fourth quarters of the same school year. If the gains made by the RMT group in the second quarter disappear with a passage of time this would mean that RMT does not have a structural conceptual impact on students' reasoning but just a temporary activation.

Figures 6.1(A) and 6.1(B) show the results of standards assessment (math benchmarks) for RMT and two comparison groups in the third and fourth quarters. The nature of the benchmarks was different in these two quarters so the comparison between classes can be made only within the same quarter. The difference between RMT and the comparison classes is the third quarter was significant (RMT vs. Comparison 1: $t = 1.8$; $p < 0.05$; RMT vs. Comparison 2: $t = 5.2$; $p < 0.0005$). In the fourth quarter the difference became even greater (RMT vs. Comparison 1: $t = 3.03$; $p < 0.005$; RMT vs. Comparison 2: $t = 5.6$; $p < 0.0005$).

We may thus conclude that RMT intervention in the first two quarters leads to a structural and self-perpetuating change in the students' conceptual understanding of the mathematical material that lasted to the end of the school year. The following vignettes illustrate students' acquisition of the rigorous approach to mathematical conceptualization and the use of cognitive functions terminology while working on the notion of fractions and operations with them.

TEACHER: Today, I want to present to you three fractions: two-thirds, one-half, and three-fourths. What might we want to do with these fractions?

STUDENT 1: We need to compare them to see which has the largest or smallest quantity.

TEACHER: That's good. Is there anything else we may want to do with them?

STUDENT 2: We may want to put them in order based on their size as quantities.

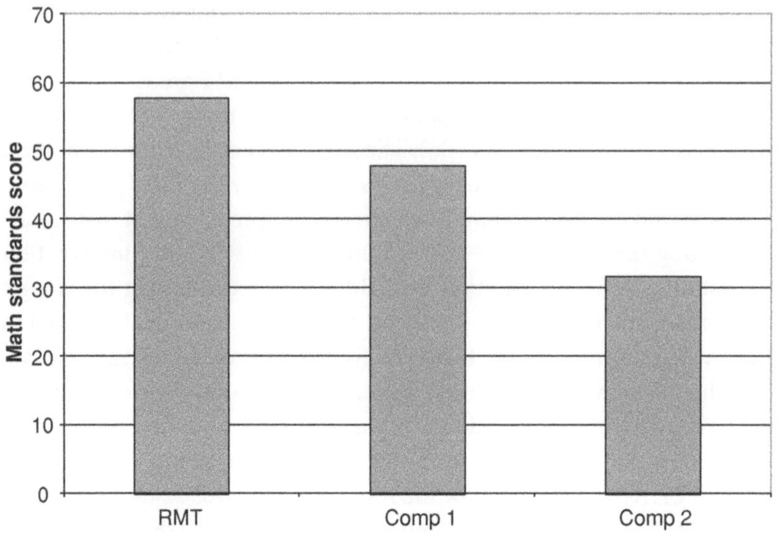

Figure 6.1(A). Results of the math benchmarks assessment of the RMT group and two comparison groups in the third quarter.

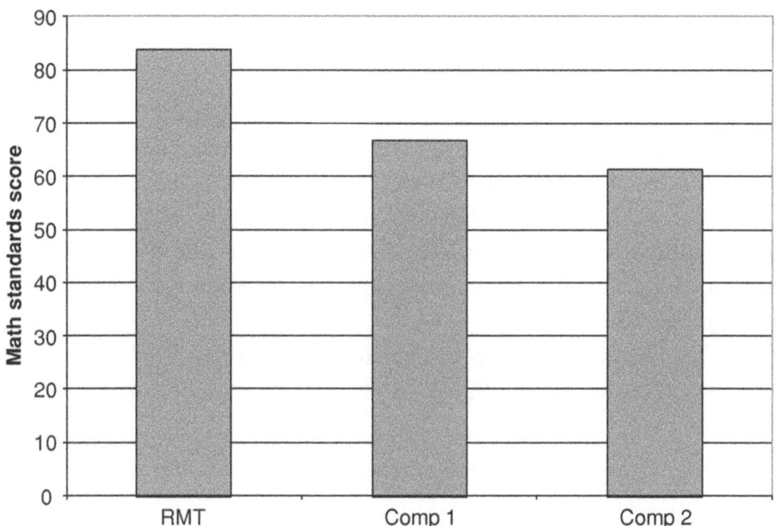

Figure 6.1(B). Results of the math benchmarks assessment of the RMT group and two comparison groups in the fourth quarter.

STUDENT 3: We might need to add them or we might want to subtract one from the other.

TEACHER: Great. Before we can do any of these what must be done first?

STUDENT 4: We have to see how they are related as quantities.

STUDENT 2: We have to form relationships between them.

TEACHER: What will help us do that?

STUDENT 5: We have to get them to speak the same language. That means we will have to translate them into the same denominator.

TEACHER: I understand that they must speak the same mathematical language. But why do we have to have the same denominators?

STUDENT 6: Well, we have to remember that each of the three fractions is a part of the same whole and that the denominator tells us the number of equal-sized parts the whole has been analyzed into.

TEACHER: That gives us some insight but I need to know more.

STUDENT 7: This means that we have to keep reanalyzing the whole in all three cases until we get the same number of equal-sized parts.

STUDENT 8: Yes. Because the numerator of each fraction tells us the number of equal-sized parts we are considering. Reanalyzing the whole changes the number of equal-sized parts we are considering.

STUDENT 4: Reanalyzing the whole changes the number of equal-sized parts we are considering without changing the value of the fraction.

STUDENT 9: Yeah. When we reanalyze the whole the numerator and the denominator will change but the fraction will still have the same value or quantity or amount.

TEACHER: This is very good. I am very surprised that none of you started talking about the lowest common multiple of the three denominators.

STUDENT 5: We could have started there but that does not help us to understand the meaning of each fraction.

STUDENT 10: The least common multiple tells us how many equal-sized parts the whole has to be analyzed into for the three fractions to communicate.

STUDENT 2: It also becomes the least common denominator.

TEACHER: So the least common multiple for the three fractions is providing us with the knowledge on how to reanalyze the whole so the fractions

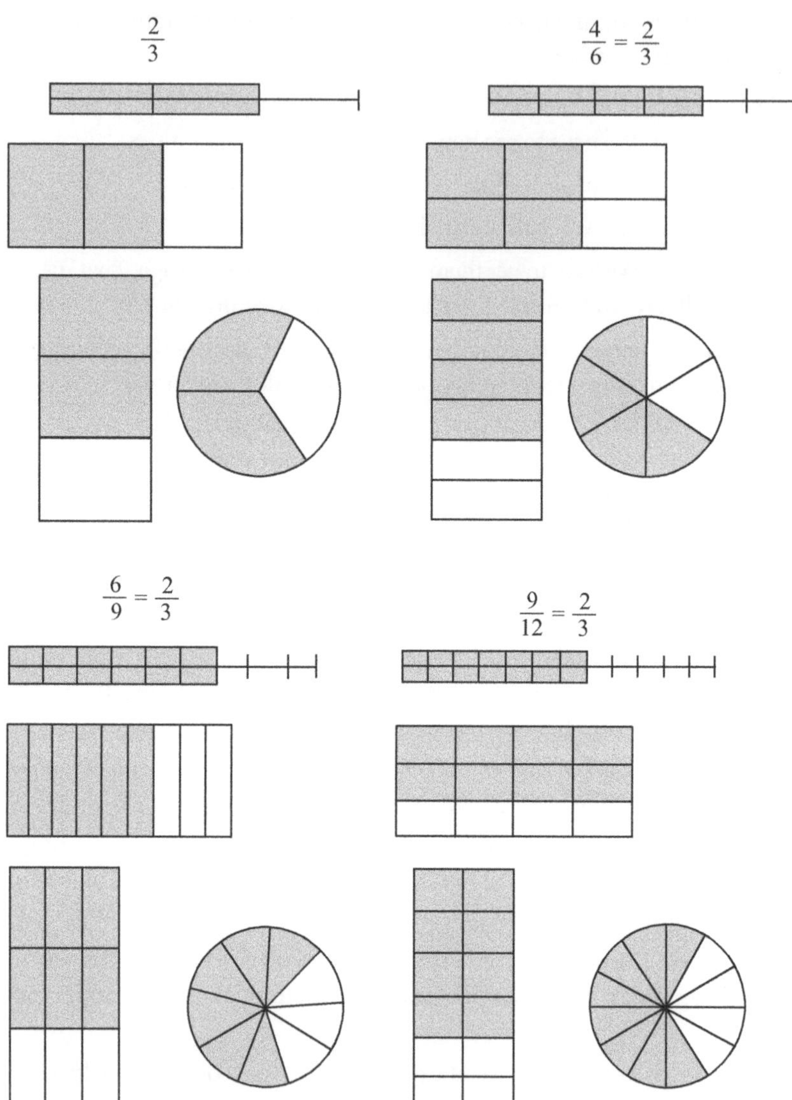

Figure 6.2(A). Representations of reanalyzing the whole for 2/3.

will have a logical basis to communicate with each other. Let us see how this helps us to change the form of each fraction so that it still keeps its value or quantity.

[Teacher mediates students to work through the process on the board. See Figures 6.2(A), 6.2(B), 6.2(C), and 6.2(D).]

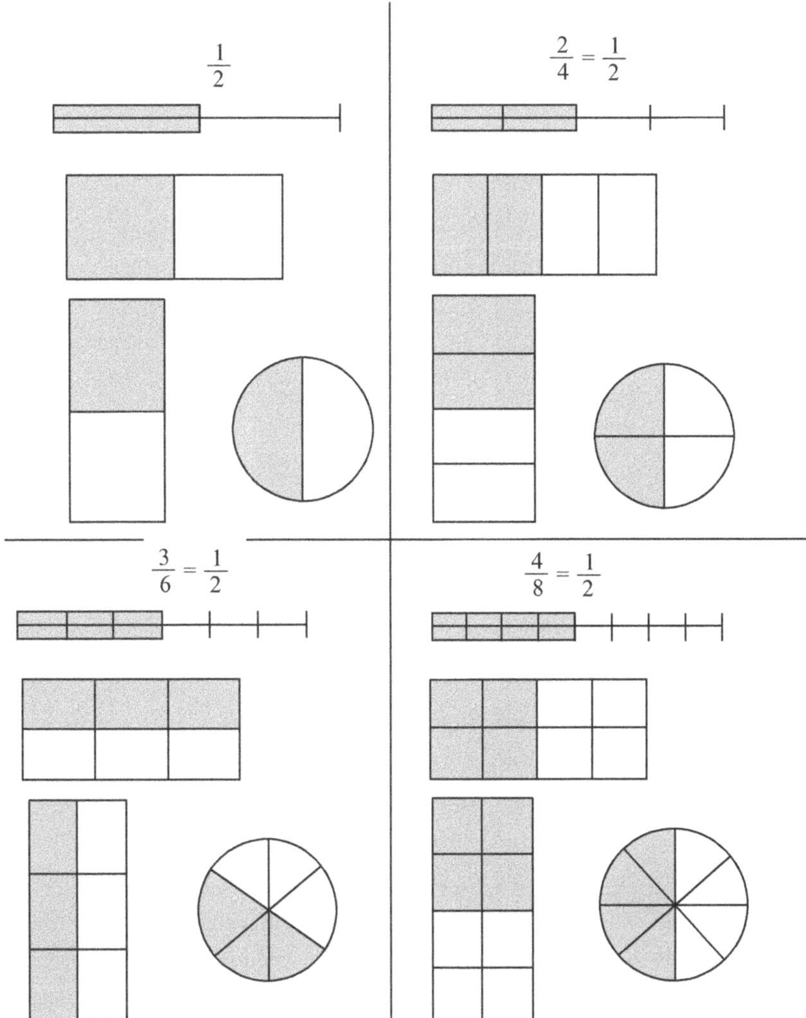

Figure 6.2(B). Representations for reanalyzing the whole for 1/2.

TEACHER: What have we accomplished by finding out that the least common multiple of the three denominators is 12?

STUDENT 11: We have important knowledge to help us to change the form of each fraction using hypothetical thinking. To convert each fraction we set up a hypothetical thinking equation.

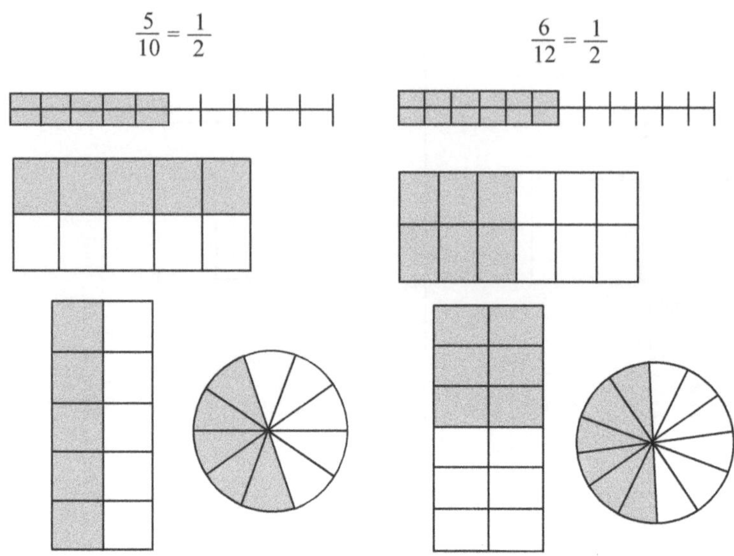

Figure 6.2(C). Representations continued for reanalyzing the whole for 1/2.

STUDENT 3: We have to conserve constancy in the value of the fraction when we set up this hypothetical thinking equation.

TEACHER: What mathematical property do we have to use as we set up this equation?

STUDENT 11: The identity property of multiplication.

STUDENT 12: Yes. When we write B/B in the equation to convert two-thirds, D/D in the equation to convert one-half, and H/H to convert three-fourths, we are using the identity property of multiplication.

TEACHER: You are telling us that the identity property of multiplication is important in conserving constancy in the value of the fraction.

STUDENT 12: It is also helping us to find the numerator in the new form of the fraction.

TEACHER: Give me your hypothetical thinking statement.

STUDENT 13: If the fraction two-thirds is going to have a new denominator twelve, as given by the least common multiple, and if three times four equals twelve, then four times two is equal to eight to give the new form of eight-twelfths.

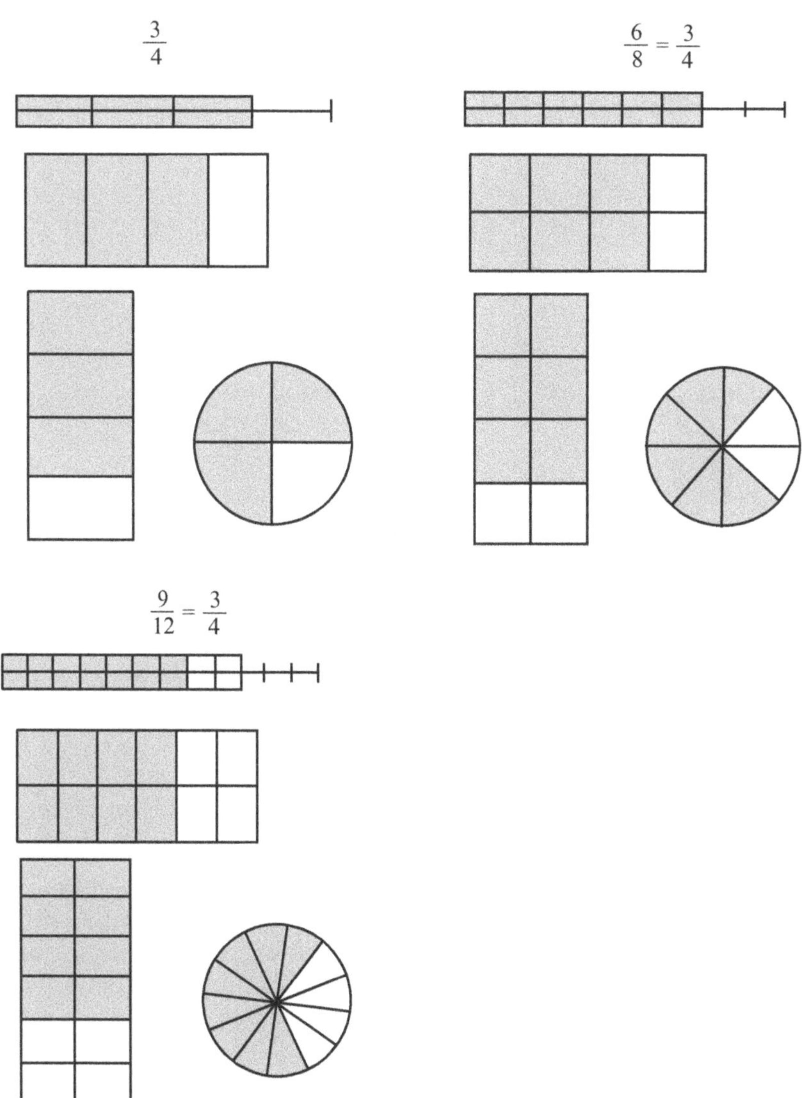

Figure 6.2(D). Representations for reanalyzing the whole for 3/4.

TEACHER: Great. Now I want you to provide more mathematical logical evidence for each conversion by using the following representations of the whole: a horizontal line segment, a vertical rectangle, a horizontal rectangle, and a circle.

Table 6.3. Providing logical evidence for fractions to communicate while maintaining their original quantities

Fraction		Multiples of denominator
2/3		3 6 9 12 15 18 21
1/2		2 4 6 8 10 12 14
3/4		4 8 12 16 20 24 28
LCM		**= 12**
Conversion of fractions		
$2/3 \times B/B = C/12$		$3 \times B = 12$, therefore $B = 4$
$2/3 \times 4/4 = C/12; C = 8$	and	$C/12 = 8/12$. Therefore, $2/3 = 8/12$
$1/2 \times D/D = A/12$		$2 \times D = 12$, therefore, $D = 6$
$1/2 \times 6/6 = A/12; A = 6$		$A/12 = 6/12$. Therefore, $1/2 = 6/12$
$3/4 \times H/H = F/12$		$4 \times H = 12$, therefore, $H = 3$
$3/4 \times 3/3 = F/12; F = 9$	and	$F/12 = 9/12$. Therefore, $3/4 = 9/12$

RMT Research with Cultural Minority Students in the Middle School

This study aimed at responding to the question of whether even a limited amount of RMT intervention aimed at the specific mathematical concept – the concept of *function* – is capable of changing students' performance.

A class of 7th-grade students at a school located in the inner city of a large metropolitan area in the Midwest received the RMT intervention for 16.5 hours over a period of 2 months. The cognitive enrichment via the IE targeted the cognitive functions prerequisite for the acquisition of the mathematical concept of *function.* All students in the RMT class ($N = 19$) were low-performing African Americans. Because this was the only 7th-grade class in the school, a 7th-grade class ($N = 18$) at a nearby school with students of the same ethnic and socioeconomic background was designated the comparison group. Students in the comparison group studied the concept of *function* by standard instructional methods and without cognitive enrichment. Prior to the start of the study, teachers of the two classes made the collective general assessment that students in the comparison group were generally more mature and, overall, had higher grades than students in the RMT group.

The cognitive pretests using the OLSAT (Otis and Lennon, 1996) indicated that the comparison group had a higher cognitive ability than the RMT group. However, the RMT group made significantly greater cognitive gains during the intervention period than the comparison group ($t = 2.23$; $p < 0.025$) so that by the end of the intervention period the cognitive performance of the two groups was similar. The effect size of the cognitive gain in the comparison group was 0.59; whereas in the RMT group it was 0.97.

The students' understanding of mathematical *function* was evaluated with the help of math concept tests (National Council of Teachers of Mathematics, 2001). At the pretest the RMT group demonstrated better results than the comparison group, though still very low in terms of their understanding the concept of *function.* The gain in the understanding of *function* by the end of the intervention period in the RMT group was significantly greater than that of the comparison group ($t = 2.7$, $p < 0.01$). The effect size of the conceptual change in the RMT group was 0.67, whereas in the comparison group there was a negative change (effect size = −0.3).

One may thus conclude that even a limited amount of RMT intervention aimed at developing the cognitive prerequisites and conceptual understanding of the notion of *function* was effective, whereas in the comparison group the traditional teaching failed to advance the students' conceptual understanding. This result replicates the finding of our previous study (Kinard and Kozulin, 2005) that used a similar research design. In that study the RMT group started at the lower conceptual level than the comparison group and yet made more significant gains. We may thus conclude that the initial level of conceptual performance is not a determining factor – students who receive RMT intervention make significant gains relative to the comparison groups irrespective of their initial level of performance.

The following vignette demonstrates the process of conceptual change experienced by the RMT group students. A major outcome from the intervention was the interaction between the enhancement of students' cognitive strategies and problem-solving skills via the IE task "Organization of Dots" (see Figure 6.3) and the emerging concept of function. The concept of constancy was first introduced as a part of the work with "Organization of Dots." Then through reflection on this work, students were mediated to move beyond the specifics of the IE task and start generalizing the concept of constancy within the contexts of dynamic change, variable, and functional relationship.

MEDIATOR: Now that you have completed this page, what changes do you observe?

STUDENT 2: The figures kept changing their positions... I mean the angles, how they are turned.

MEDIATOR: Could we say that the orientation of the figures changed?

STUDENT 11: Yes, that's a good word.

STUDENT 2: Yeah. That's what I'm saying.

STUDENT 10: Well, I noticed that the figures got more entangled.

STUDENT 12: Yes. There is greater intrusion of figures into each other's space.

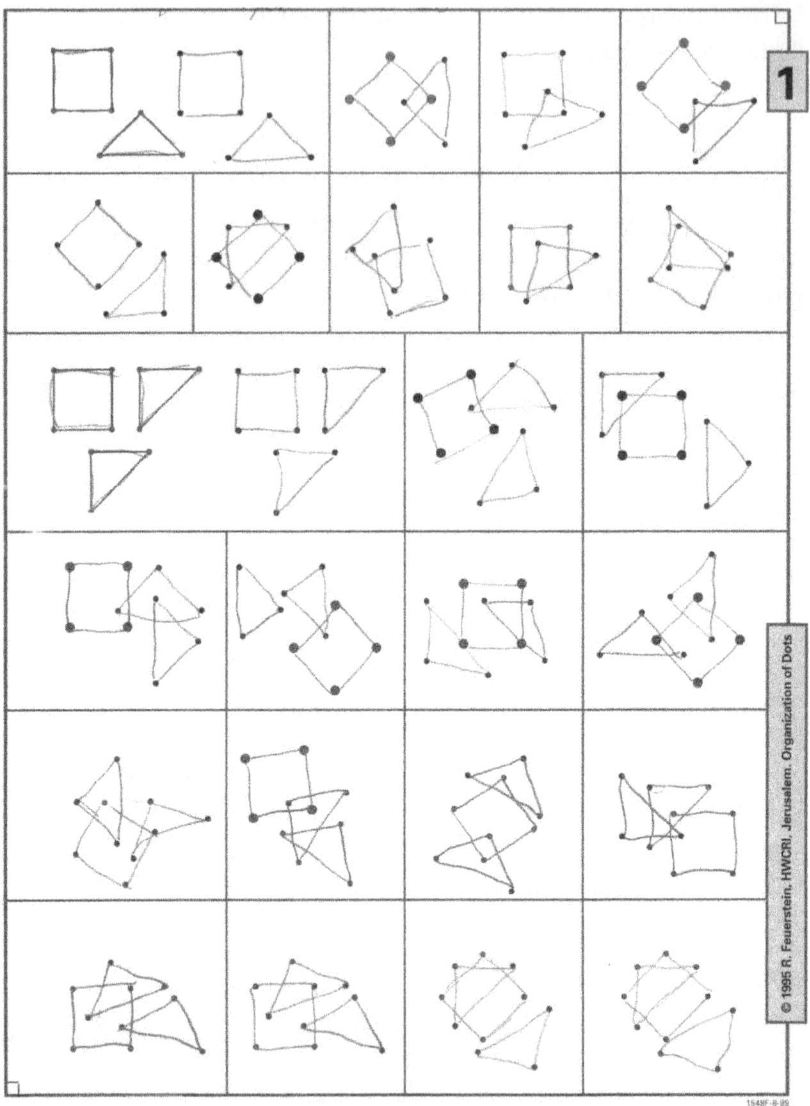

Figure 6.3. Student's work with "Organization of Dots" tasks.

STUDENT 7: I agree. I would say that the figures overlap more as we move down the page.

STUDENT 14: The dots keep getting closer together as we move down the page, too.

MEDIATOR: I hear you saying that the orientation of the figures, the closeness of the dots, and the overlapping of figures are changing as we go down the page. Because these things are changing what are they?

[Pause.]

MEDIATOR: Well, are they staying the same? Are they constant?

STUDENT 1: No! We just said that they are changing.

MEDIATOR: So when something changes it does what?

STUDENT 4: It expands.

STUDENT 11: It could shrink.

MEDIATOR: So what is a word that covers both expanding and shrinking?

STUDENT 5: It transfers?

STUDENT 11: It modifies?

MEDIATOR: Good! Now what is another word that means it changes or it modifies?

STUDENT 2: I know. It varies.

MEDIATOR: Great! So something that varies is a what?

SEVERAL STUDENTS [almost in unison]: It's a variable!

MEDIATOR: What are the variables here?

STUDENT 7: The variables are the closeness of the dots and the overlapping of the figures.

MEDIATOR: That's correct.

STUDENT 15: You know, I think there is a functional relationship between the closeness of the dots and the overlapping of the figures.

MEDIATOR: That's powerful. Let us call the closeness of the dots the proximity of the dots. Is the proximity increasing or decreasing as we move down the page?

STUDENT 3: The proximity is increasing.

MEDIATOR: What about the distance between the dots?

STUDENT 9: The distance from dot to dot is decreasing as we move down the page.

STUDENT 15: So there is a functional relationship between the proximity of the dots and the overlapping of figures.

STUDENT 3: The proximity of the dots is the controlling variable. It is the independent variable. The overlapping of figures is the dependent variable.

MEDIATOR: That is terrific!

STUDENT 6: I see layers of functional relationships.

MEDIATOR: Tell us more.

STUDENT 6: We organized the loose dots onto functional relationships using the psychological tools. These tools function the way they do because of their structure and the functional relationships that build their structures.

STUDENT 4: Now we have created a new functional relationship between the proximity of the dots and the overlapping of figures.

STUDENT 12: Yeah. I think there is another functioning going on. We have been networking our cognitive functions to help us comprehend all that is happening.

MEDIATOR: This is fantastic. How do these layers of functional relationships differ? Is one more abstract than the others?

STUDENT 10: The functional relationship coming from the networking of the proximity of the dots with the overlapping of figures is more abstract than the psychological tools.

STUDENT 8: The networking of the cognitive functions is the most abstract.

STUDENT 3: I agree. Using these cognitive functions is a powerful tool. Can we call this a psychological tool?

STUDENT 5: It sure has me thinking at a high level.

At this point the mediator asked the students were there functional relationships in real life. In response one student stated:

> I woke up this morning and it was snowing, real big, fluffy flakes. It's been snowing since midnight. The lady who gave the weather forecast on TV said there was a good chance for snow because of what the temperature and relative humidity were going to be. This was around 6:30 in the evening and I had no idea it was going to snow. But sure enough it snowed, and it's still snowing. I think there is a functional relationship between temperature, humidity, and the chance of snowing. The pressure might have something to do with it too. I'm not sure.

High School Dropouts

In this section we present the results of a RMT approach that goes beyond the typical classroom situation and reaches out for high school dropouts with a history of systematic unemployment (see Kinard, 2001). The RMT implementation was a part of a larger project aimed at equipping these "high-risk" young people with useful job skills while at the same time revitalizing dilapidated housing and reducing toxic environmental contaminants from inner-city communities.

In the first pilot project we tested the hypothesis that integration of IE lessons into the training process will positively affect the trainees' cognitive skills, their job-related motivation, and their success in passing the end of the course exams related to the techniques of decontamination of polluted environments. When embarking on this project we were not sure that the trainees would accept the IE component that for an outsider may look rather abstract and not directly related to her life and employment needs. Nevertheless trainees themselves apparently were quite capable of grasping the meaning of cognitive skills for their life and future. This is how one of them conveyed his impressions from the work with "Analytic Perception" IE tasks in his reflections journal:

This instrument basically assisted me in analyzing difference between situations or individuals by finding underlying problems within a problem, i.e. the natural root or cause, when the surface or distractionary problem at hand is superficial. Breaking down the component or problem (dissecting or systematically searching) and rebuilding from there may resolve what is at hand and also clear all other surfacing or surfaced directions that are the hidden root or cause.

An independent observer who watched the sessions commented:

What I have observed from the men in the Hazardous Waste Management class is remarkable. Their whole outlook on life has changed for better. It seems to me that they have a sense of urgency about getting started and succeeding in the goals they have set forth. With a lot of the negativity in the East Palo Alto, it is great to see a course like this change the lives of these young men, and I hope this course will continue to change the lives of many more.

Probably the best testimony of the change in the trainees' habits of heart and mind is the following comments made by a wife of one of the students:

My husband is in the class and I see a tremendous change in him. Before this class started he had jobs but they did not last more than six months. I told him he should go to school and get some type of training to get a better job, but he always ignored me. I guess he changed his mind and he went to school. He cannot stop talking about the class and how he wants to start his own business. This class changed him so much that after this class is finished he is going to take a business class. I am very proud of him for going back to school. This class is the best class that I [have] seen in a long time. They are there everyday; no one misses the class unless it is an emergency. They respect everyone. I always see a smile on everyone's face, especially when they passed their test. That was the happiest group of all of them.

These and similar comments and reflections testify to a profound change in the trainees' learning and job habits. One may ask, however, whether the program affected exclusively the participants' self-perception and motivation or that it also changed their cognitive and performance skills. The last question was answered by comparing the pre- and postprogram results of the RL-3 logical reasoning test (Ekstrom, French, Harman, and Dermen, 1976; Ekstrom, French, and Harman, 1979) and the results of the technical exams given at the end of the training program. The RL-3 test evaluates the students' ability to reason from premises to conclusions. It also tests the attention students give to essential details while ignoring irrelevant information. This is a verbal test that requires considerable text comprehension effort; at the same time, the program had no units specifically targeting reading skills. The improvement in RL-3 results may thus be considered an example of a far-transfer of cognitive skills developed through IE to the area of verbal thought.

Different groups of students received different amounts of IE training ranging from as little as 15 to as much as 72 hours. In the majority of groups the pre- to postgain in the cognitive RL-3 test was significant. To determine whether the improvement in logical reasoning skills on average had an impact on trainees' achievement on the end-of-program technical examination we pooled all the RL-3 posttest results and correlated them with the results of the three technical exams given at the end of the program. The Pearson correlations were significant for all three exams, ranging from 0.79 to 0.74 ($p < 0.01$). It is important that the correlation between RL-3 pretest results and the final exam results was nonsignificant. These results thus allow us to be optimistic regarding the possible benefits of cognitive enrichment with a population of "high-risk" young adults.

The next question is whether similar populations of "high-risk" learners would demonstrate improvement in their mathematical and science skills when exposed to the RMT intervention. As in the previous example the RMT intervention was part of a larger 6-month training aimed at equipping unemployed, "high-risk" inner-city residents with construction and environmental remediation skills. An extensive evidence of conceptual change in mathematics and science was documented through chronicled student work and videotaped sessions. Students engaged in the full cycle of mathematical investigation – representation, manipulation, and validation – producing their

Table 6.4. Time and distance object has fallen from spaceship

Time object has fallen (seconds)	Distance object has fallen (feet)
1	5
2	31
3	76
4	140
5	223
6	325
7	446

own mathematical models and deepening their understanding of velocity, acceleration, gravity, force (including centripetal and centrifugal), the notion of relativity, and so on. In separate applications students demonstrated that they can derive structure/function relationships and functional relationships between variables during field investigations at a local planetarium, science and industry museum, natural history museum, butterfly haven, aquarium, and linear accelerator.

The following vignette illustrates the students' reasoning near the end of the 76 hours of RMT intervention provided over a 3-month period. The RMT intervention included a rich battery of IE-based cognitive tasks using different modalities together with mathematically specific tools that focused on teaching algebraic concepts, including mathematical function. Students had previously performed advanced cognitive tasks dealing with numerical progressions, relationships within progressions, and relationships between relationships. It was at this point that they were introduced to a number of physics concepts such as inertia, energy, force, momentum, and relativity of time and space.

First the class was issued the following information and data:

Consider an object with a mass of 2 pounds dropped from a spaceship in proximity to the surface of a planet with no atmosphere. Gravitational pull on the object causes it to fall according to the following data (Table 6.4).

Without any further information or instructions students immediately began to individually and in small groups form two coinciding progressions. Shortly thereafter, they collaborated and placed the data shown in Figure 6.4 on the board.

TEACHER: What is this you have constructed on the board?

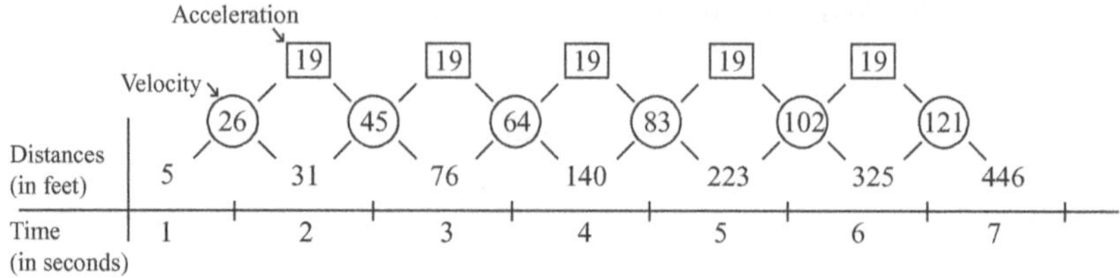

Figure 6.4. Progressions of distance and time.

STUDENT 1: The first progression at the bottom goes from left to right. It is the time in seconds that the object has fallen from the spaceship. The second progression is the total feet the object has fallen from the spaceship.

STUDENT 2: We are looking at everything by going from left to right.

STUDENT 3: The numbers in the circles show the relationships between the distances the object has fallen. The first number is twenty-six. This is the number of feet the object falls from one second to two seconds. It is also the distance from five feet to thirty-one feet.

STUDENT 4: The number in the second circle is the distance the object falls from two seconds to three seconds. It is also the distance the object falls from thirty-one feet to seventy-six feet.

STUDENT 5: These relationships form a progression also.

STUDENT 6: Each relationship is the velocity of the object.

TEACHER: Explain.

STUDENT 6: Well, the object is falling toward the surface of the planet. We learned earlier that the velocity is the distance of movement per unit of time in a certain direction. From five feet to thirty-one feet, the velocity of the object is twenty-six feet per second. From thirty-one feet to seventy-six feet, the object's velocity is forty-five feet per second.

STUDENT 7: So the velocity of the object is changing and is a variable. The distance the object has fallen and the time it has fallen are also variables.

STUDENT 8: There is a functional relationship among the two variables, the time the object has fallen, and the distance the object has fallen.

STUDENT 3: There is also a functional relationship between the distance the object has fallen and the velocity of the object.

TEACHER: What is occurring at the next level in your diagram?

STUDENT 2: Here we have the relationship between relationships.

STUDENT 9: We also can say there is a functional relationship between the time the object is falling and the object's velocity.

STUDENT 10: Here we are seeing how the velocity of the object is changing over time. The velocity is in feet per second. This is how the distance of the object changes with time. Now we are seeing how the velocity of the object is changing with time. This is the feet the object falls per second per second.

STUDENT 11: This is the acceleration of the object as it falls toward the planet. The object is accelerating nineteen feet per second per second.

STUDENT 1: The acceleration is the change in the object's velocity per unit time.

TEACHER: Could you explain this in a different way?

STUDENT 13: Yes. Each second that passes the object increases its speed toward the planet by nineteen feet per second.

STUDENT 8: The acceleration of the object toward the planet is conserving constancy. It is uniform over the distance and time the object is moving.

TEACHER: Where is this acceleration coming from?

STUDENT 14: It is coming from the gravitational pull the planet is exerting on the object toward its surface at nineteen feet per second per second.

TEACHER: Is there any significance that the object has a mass of two pounds?

STUDENT 5: Activating my prior knowledge, we learned that force is the pull or push on an object of a certain mass to accelerate its motion or to change its velocity. This is the force it takes to move this two-pound object nineteen feet per second per second. This means that the force on the object is thirty-eight pounds per feet per second per second.

TEACHER: Could you call the presentation (see Figure 6.4) you have constructed on the board a system of psychological tools?

STUDENT 11: Sure. We used the data given to us on the object falling from the spaceship to form these progressions, relationships, and relationships between relationships that caused us to use many cognitive functions and helped us to more deeply understand force.

STUDENT 7: We have been building our understanding around the formula $F = ma$ or force equals mass times acceleration. This formula and the psychological tools on the board have helped us to integrate cognitive functions like forming proportional quantitative relationships, projecting and restructuring relationships, quantifying space and spatial relationships, quantifying time and temporal relationships.

STUDENT 2: As we analyzed and integrated the data about the falling object we were able to construct and use the tools of the progressions. This required us to use those cognitive functions you mentioned and others. The relationships among time, distance, velocity, acceleration, mass, and force connect with the formula $F = ma$. This has really helped me to develop a deeper understanding of force.

STUDENT 14: While we were talking I thought about using the mathematical psychological tool of the x-y coordinate plane to graph the functional relationship between time the object has fallen and the distance the object the object has fallen (see Figure 6.5) and time the object has fallen and the velocity of object (see Figure 6.6).

TEACHER: Will you put them on the board please?

[Student draws graphs on the board.]

STUDENT 5: The functional relationship between the time the object has fallen and the distance the object has fallen is not linear.

STUDENT 9: The unit functional relationships or the slopes are changing.

STUDENT 14: Yes. You can see that as time increases the unit functional relationship also increases. This unit functional relationship is the velocity.

TEACHER: Great. That is powerful.

STUDENT 7: The functional relationship between the time the object has fallen and the velocity of the object is linear. The unit functional relationship is the acceleration.

STUDENT 10: As we said before the acceleration of the object conserves constancy.

TEACHER: This is very good work.

Test results of cognitive and math performance collected in one of the groups of "high-risk" students ($N = 5$) indicated that they indeed made significant gains in both areas. Cognitive gains were measured by the OLSAT (Otis and Lennon, 1996), whereas the students' understanding of mathematical *function* was evaluated with the help of math concept tests (National Council of Teachers of Mathematics, 2001). The effect sizes of both cognitive and math achievement were large: 1.3 and 1.9, respectively. These results confirm that at least some "high-risk" learners have sufficient learning potential for benefiting from RMT intervention. The emphasis here is on potential.

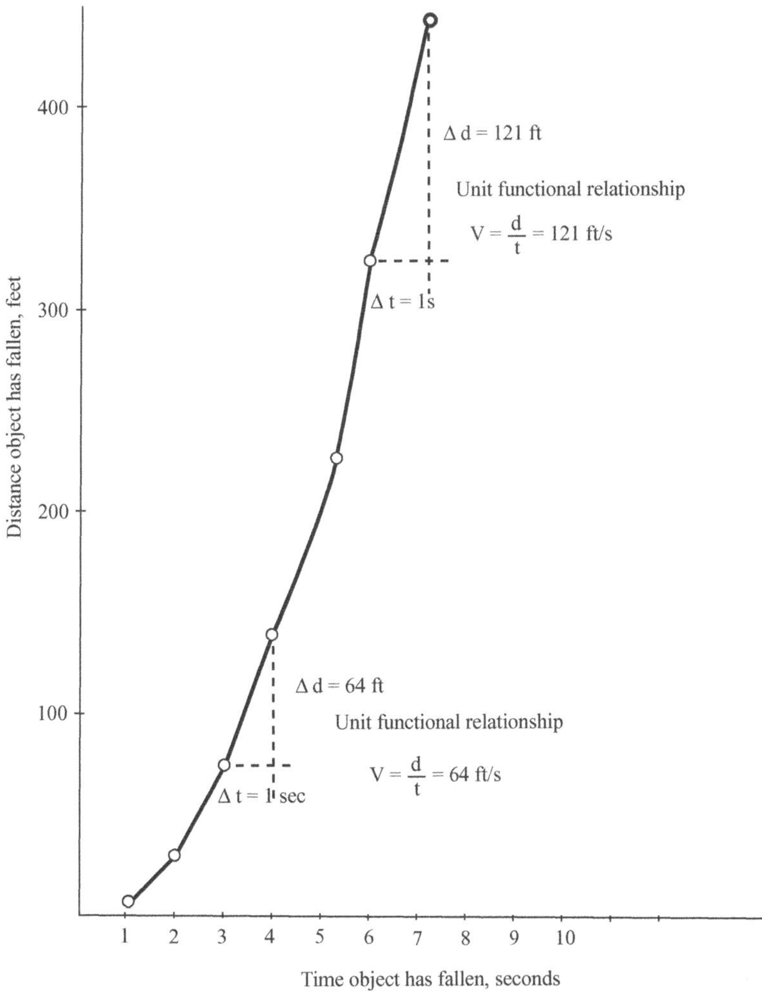

Figure 6.5. Functional relationship of velocity and time object has fallen.

Only a large-scale and well-controlled study may answer the question regarding the effectiveness and generalizability of the RMT approach with high school dropouts. However, even the present findings seem to be important in spite of the small numbers, taking into account a still popular belief that individual cognitive performance is a rather stable trait resistant to significant improvement in adults (Herrenstein and Murrey, 1994). In our opinion the fact that at least some of the "high-risk" young adults are capable of dramatic change in their cognitive performance and conceptual reasoning already justifies an educational effort. As the case of Stephen Hawking is sufficient to prove

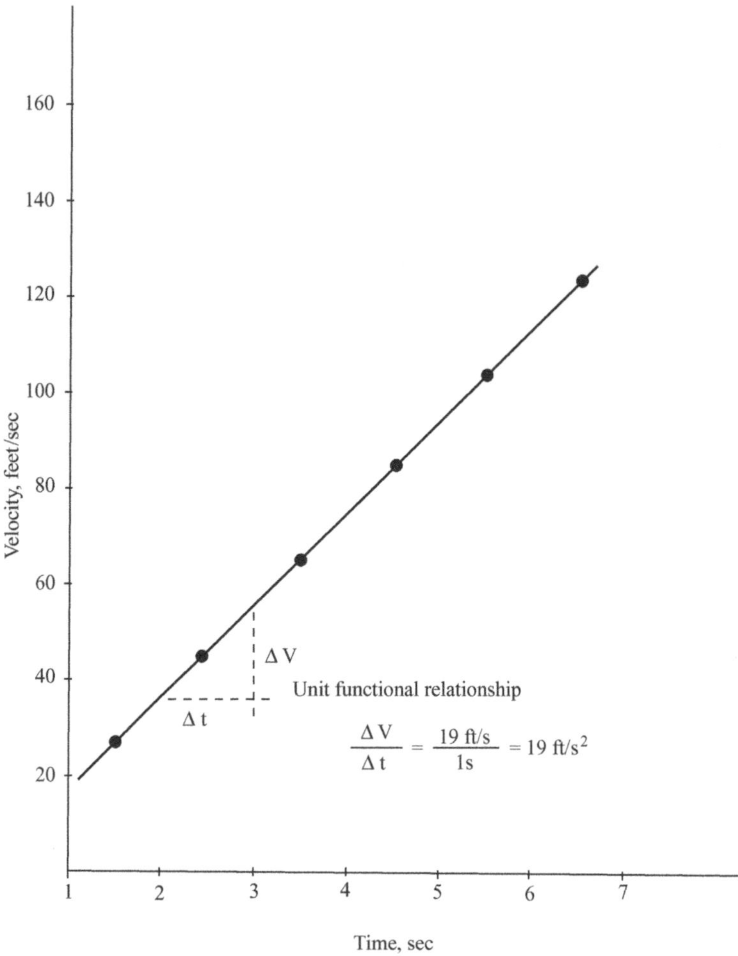

Figure 6.6. Functional relationship between velocity of object and time.

the *possibility* for a severely handicapped person to become a brilliant scientist, a few cases of significant cognitive and conceptual changes in "high-risk" adult learners proves that this population is potentially capable of reaching far beyond its initial performance level.

Teacher Change

In this section we examine changes observed in teachers who received RMT training and become involved in RMT classroom practice. The following areas were examined: teachers' beliefs, instructional delivery, views of mathematics curricular, student learning, teachers' lesson plans and their planning process, how teachers analyzed student work, and how teachers valued student oral and written responses. The evidence of change was obtained during the pre- and postprogram meetings with teachers and from their reflective diaries.

Before RMT training all teachers involved expressed serious doubts regarding the need of teaching students cognitive processes, let alone insisting that students identify and articulate the meaning of cognitive functions. In addition, the teachers were convinced that students who failed to grasp basic arithmetic skills early in their education would never be able to achieve in mathematics. The teachers' perception of mathematical learning focused on the need for repetition and the "drill and kill" approach. After RMT training and classroom practice these same teachers became vigorous supporters of the need for explicit teaching of cognition. They became convinced that students should be able to identify and articulate the meaning of their own cognitive functions and that such an identification should be an ongoing process throughout each class session.

Before RMT training most teachers' instructional delivery was based on the frontal teacher-centered model. Following training and classroom practice the teachers' instructional delivery model shifted significantly in the direction of acknowledging students as active agents of learning. This brought a change from lecturing, giving examples, and assigning tasks similar to the teacher's examples to mediating for cognitive development and conceptual learning. To a considerable extent mechanical and algorithmic methods became replaced by a cognitive-conceptual constructive approach.

The change was also observed in teachers' perception of the mathematics curriculum and its relation to lesson plans and the lesson planning process. Instead of accepting curricular material as a "given" presented to them by administrators, teachers began to search for and identify the "big ideas." They started performing active structural analyses of the mathematics content, forming relationships between the concepts and systematically reorganizing the concepts from the big idea to the supporting subconcepts. At this point teachers began to analyze each concept in terms of the prerequisite cognitive functions essential for helping students' conceptual understanding. This led teachers to determine which cognitive tasks would be used to develop or mobilize the cognitive functions and which general psychological tools are needed to be appropriated by students to perform the tasks. Teachers also identified which mathematically specific psychological tools would be needed in the process and performed an operational analysis to determine how the selected cognitive functions would be orchestrated by each mathematically specific psychological tool to build the conceptual understanding. This process provided not only a framework and guideline for lesson planning but also a mechanism for RMT classroom instruction and learning. Prior to the RMT training the notion of psychological tools, general and mathematically specific, was not a part of their repertoire. Mathematically specific psychological tools existed only incidentally as forms of mathematical content, without structure/function relationship and instrumentality. After the training they became an integral part of their didactic approach.

Below are some examples of teachers' reflection on the change they experienced in the course of RMT training and classroom practice.

Teacher Comments

Teacher A Comments (before RMT Training)

Students were taught to memorize and give answers without explanation. Most students couldn't explain how or why they got their answer. They didn't have the processes or reasoning for the math skills they used.

I used to just show, allowing students minimal time to understand concepts completely. They weren't expected to give answers or really have an opportunity to understand the concept internally.

Teacher A Comments (after RMT Training and Practice)

They [students] take more ownership for their answers. They are able to support or explain the steps in solving math problems.

Conceptual learning describes the math concepts learned through the use and understanding of the cognitive functions. Using psychological tools enhances the learning process of a concept.

Teacher B Comments (before RMT Training)

I believed lecturing and modeling were the most effective ways to teach students math concepts. Using manipulatives and showing how to solve problems was the only way I knew. I had difficulties myself explaining how I worked through a process because I was unaware of my own thinking; therefore, it was difficult to teach my students how to explain their thinking.

Teacher B Comments (after RMT Training and Practice)

Having the students be aware of their thought process has become so critical. Using the cognitive functions as tools to guide them through math concepts has made the understanding of the process more effective. I teach less skill and drill and more understanding of the process in the concepts. It takes longer, but because it's more meaningful to the students, it really sticks with them. My teaching has also changed in the way that there is less lecturing and more discussion! Discussion of the why's and how's has become key in the understanding.

Teacher C Comments (before RMT Training)

I believed that if I modeled and had some hands on activities (which *do* help), my students would understand the concepts I was trying to get them to learn.

Teacher C Comments (after RMT Training and Practice)

I now believe students need to DEEPLY understand the concepts in math in order to learn mathematics. This needs to happen through my being a mediator instead of a 'modeler.' I do still use hands on activities to give children a chance to use manipulatives. This usually happens **after** they have had the concept mediated to them.

I have gone from lecturing and modeling to **mediating**. Instead of **telling** them how to do problems, concepts in math, I now scaffold and mediate my students through the cognitive functions to get to the concepts so they have a deep understanding of what they need to know.

Conceptual learning in math means that we go back to the six core concepts. I build the concepts through the cognitive functions and math specific psychological tools. For example, when I am teaching about variables, I can use the mathematically specific psychological tool of equations/formulas. The core concept I am building to and teaching is quantity.

Teacher D Comments (before RMT Training)

My belief was that students learn math by observing the teacher solve a couple of problems on the board. If this strategy were done enough times, then the students would have been able to master the concept.

Instruction delivery: I model the procedures for the students hoping that they would then look at their notes when faced with similar problems. This teaching practice resembles the way I was taught.

There was little work analysis. My main focus was 'the students either get right or wrong' I paid no attention to the students' thinking process.

Oral and written responses had little value to me. I used to say, 'if you know how to solve the problem, you have to show me on paper.' I had little interest in what the students had to say or write about a problem.

Teacher D Comments (after RMT Training and Practice)

Students need to be interested in order to do "work." Here is where I, as a mediator of Knowledge, need to have the intentionality to get the students' attention. Students learn math by forming relationships with their own lives. Their prior knowledge becomes crucial at this point. The beauty of it is that every single student has some prior knowledge about certain topics.

Instruction delivery: I have to start by saying that my vocabulary changed. I began to use terms like, activating prior knowledge, forming relationships, conserving constancy, and 'just a moment, let me think.' More attention was paid to the thinking process rather than the final answer. I used more pictures and real life examples to capture the attention of the kids. This way a math lesson became alive because it had a connection with the lives of the students.

We spend more time analyzing (whole class discussion) student work. The students were active participants in their own learning. We learned that making mistakes is a natural way of learning. At the same time, we were in the process of becoming better learners.

I truly valued student oral and written responses because it was a chance for me and for the students to celebrate the tremendous progress they were achieving. It was a melody to my ear to hear students say to each other, 'you need to provide your logical evidence' or 'I have a connection to make.'

Conclusion

We started our quest for rigorous mathematical thinking with a question of what constitutes the core of mathematical culture and how this culture can be appropriated and internalized by students in the mathematics classroom. A brief review of various reform attempts made since the 1960s demonstrates that in spite of good intentions of many reformers mathematics education in the United States is still suffering from three major problems: the standards of mathematics education are set in terms of products rather than processes, the cognitive prerequisites of efficient mathematical learning do not receive proper attention, and the activities taking place in mathematical classrooms rarely foster the development of students' reflective and rigorous mathematical reasoning. In this sense many of these activities do not qualify to be called "learning activities" in a proper sense because they do not lead students toward becoming self-regulated and independent learners. Though all students suffer from the above-mentioned educational inadequacy, those from socially underprivileged groups suffer more because their immediate environment does not offer compensatory mechanisms available to more privileged students.

We then proposed a blueprint for the development of the Rigorous Mathematical Thinking (RMT) paradigm based on a double theoretical foundation of Vygotsky's (1986) concept of psychological tools and Feuerstein's (1990) notion of mediated learning experience. Vygotsky envisaged human cognitive functions as shaped by the sociocultural tools and interactions available in given society. Symbolic mediators, such as signs, symbols, texts, formulae, pictures, and so on, constitute a potentially powerful stock of tools that can organize and shape human attention, perception, memory, and problem solving. Once internalized these tools become inner psychological tools of a person. The proposed RMT paradigm uses Feuerstein et al.'s (1980, 2002) Instrumental Enrichment program to expand and enrich the students' system of general psychological tools and in this way enhance their cognitive functions. Further, following the logic of Vygotsky's theory we extended the notion of psychological tools toward such specific mathematical tools as number line, data table, the *x-y* coordinate system, and graphs. One of the primary tasks of mathematics education according to the RMT paradigm is to create conditions for students' appropriation of these symbolic systems as tools that can shape the process of their mathematical reasoning and concept formation. In other words, it is not enough to provide a student with, for example, the *x-y* coordinate system as an external tool for manipulation with specific data; the aim is to facilitate the process of internalization so that the concept of the *x-y* system becomes an inner psychological tool for students' thinking about all kinds of data.

The theory of mediated learning experience (Feuerstein, 1990) provides us with a model of teacher/student interaction. Instead of being just a source of information and rules, the teacher in this model is expected to be sensitive to the cognitive needs of the students and to shape the teaching/learning interaction in a way that fosters cognitive functions and problem-solving strategies. The concept of mediated learning together with the notion of mathematically specific psychological tools was then translated into specific didactics of RMT teaching and teacher training. The "mechanics" of RMT classroom teaching and learning interactions was illustrated by the vignettes derived from actual classroom cases. In addition, several cases of RMT applications with different groups of underprivileged students were systematically researched and the pretest, posttest, and follow-up data analyzed to demonstrate the effectiveness of the RMT approach. We put particular emphasis on the potential for rigorous mathematical reasoning demonstrated by students belonging to severely disadvantaged groups. The very fact that such potential does exist and can be identified justifies the investment of instructional time and teachers' energy into the cases that otherwise might be considered hopeless.

Now it is time to focus on still unresolved issues and future directions for research and implementation. The first one is the issue of assessment. Our claim is that the current emphasis on summative assessment that focuses on students' mastery of certain mathematical rules and operations should be complemented by at least an equally strong emphasis on formative assessment. While summative assessment responds to the question, "What the students can do now," formative assessment helps to formulate the intervention strategy "What should be done to help the student to achieve the instructional objective." In our opinion the most promising perspective for formative assessment is so-called dynamic or learning potential assessment (see Haywood and Lidz, 2007; Sternberg and Grigorenko, 2002). Instead of checking what children can do at a present moment, dynamic assessment focuses on what Vygotsky (1986) calls their zone of proximal development. Dynamic assessment thus helps to distinguish between the present performance level and the learning potential of the child. Moreover, it also identifies those intervention strategies that demonstrate their efficiency in helping children to construct new psychological functions and operations. Cognitive psychology and education already accumulated considerable experience in dynamic assessment of such cognitive functions, as perception, attention, memory, and general problem solving. What is still missing is a strong curriculum-based dynamic assessment. This task will remain complicated as long as mathematics standards are perceived in terms of content. If, however, we approach mathematical standards from the perspective of concept formation, then the dynamic assessment of mathematical reasoning is quite attainable. The amount of work, however, is not going to be small because for each one of the conceptually understood standards a dynamic assessment procedure should be developed and a form of mediation given during the assessment should be elaborated. The benefits, however, may be enormous, because instead of a simple dichotomy of achieving and underachieving students, we would be able to identify students both in terms of their achievement level and their learning potential and, in addition, receive information regarding those forms of mediation that are particularly effective for enhancing the learning abilities of a given student.

The second issue is mediation provided by the teachers. One of the central themes of the RMT paradigm is the need to turn the mathematics teacher from being a mere provider of information and rules into a mediator. One of the central differences between these two roles is that the teacher-mediator does not take the students' cognitive functions for granted but actively explores the cognitive status of his or her students and shapes the instructional situation in such a way that it promotes the students' cognitive development (on current approaches to teaching thinking, see Harpaz, 2007). The teacher-mediator should also be well versed in the conceptual aspects of mathematical culture and be able to teach mathematics conceptually rather than just procedurally. Last, but not least, the teacher-mediator should be able to engage students in active interactional learning activities in the classroom without sacrificing the requirement for mathematical rigor.

These tasks require certain changes in both preservice and in-service training of mathematics teachers. First, we strongly believe that preservice training of future mathematics teachers should include a serious study of cognitive processes not only theoretically but also practically. It would be desirable to include training in some of the cognitive enrichment programs, for example,

Feuerstein et al.'s (1980) Instrumental Enrichment, in the college curriculum so that future teachers receive a hands-on experience in the development of their own and their students' cognitive functions. Experience with cognitive enrichment programs may also serve as a platform for teaching the techniques of mediation. In addition, mathematics teachers should receive a deeper understanding of mathematics culture and the conceptual aspects of mathematical knowledge. There is no justification for the United States to be so severely behind other countries in the percentage of primary school teachers who received specialization in teaching mathematics (Ginsburg, Cooke, Leinwand, Noell, and Pollock, 2005). Finally, in-service training should include additional experience in teaching mathematics to students who demonstrate serious cognitive problems, either because of a learning disability or sociocultural disadvantage. The latter objective will inevitably come to the fore with more and more special needs children being integrated into regular classrooms.

The third issue is the need for redesigned instructional materials. The objectives of the RMT approach require text- and workbook materials that respond to the need for cognitively oriented, conceptual, and mediated learning and teaching. In this respect the steps already taken in adapting the Vygotskian mathematics curriculum to the needs of the primary school in the United States (Davydov, Gorbov, Mikulina, and Saveleva, 1999) should be continued and extended. At the present moment these materials cover only the first few years of the primary school. The materials compatible with RMT should be constructed for middle and high school. Moreover, special emphasis should be made on the development of remedial materials aimed at students who for a variety of reasons reached high school without cognitive functions and mathematical conceptual knowledge appropriate for the challenges of the high school mathematics curriculum. The new materials should assist mathematics teachers in turning their lessons into a genuine learning activity that leads the students toward becoming reflective, self-regulated, and, ultimately, independent learners.

References

Alexandrova, E. I. (1998). *Matematika: Uchebnik dlia 1 klassa* (Mathematics for the 1st grade). Moscow: Dom Pedagogiki.

American Association for the Advancement of Science. (1990). *Science for all Americans: Project 2061.* New York: Oxford University Press.

American Association for the Advancement of Science. (1993). *Benchmarks for Science Literacy: Project 2061.* New York: Oxford University Press.

Anderson, N. (1996). *A functional theory of cognition.* Mahwah, NJ: Lawrence Erlbaum.

Atahanov, R. (2000). *Matematicheskoe myshlenie I metodiki opredelenija urovnej ego razvitija* (Mathematical reasoning and methods of the assessment of the stages of its development). Riga, Latvia: Eksperiments.

Boyer, C. B. (2004). *History of analytic geometry.* Mineola, NY: Dover Press.

Boyer, C., and Merzbach, U. (1991). *A history of mathematics.* New York: Wiley.

Brown, A., and Ferrara, R. (1985). Diagnosing zones of proximal development. In J. Wertsch (Ed.), *Culture, communication and cognition: Vygotskian perspectives* (pp. 273–305). New York: Cambridge University Press.

Bruner, J. (1968). *Toward a theory of instruction.* New York: Norton.

Cajori, F. (1993). *A history of mathematical notations.* New York: Dover.

Campbell, G. (Winter 1999/2000). Support them and they will come. *Issues in Science and Technology Online.* Retrieved January 4, 2005, from www.issues.org

Campione, J., and Brown, A. (1987). Linking dynamic assessment with school achievement. In C. Lidz (Ed.), *Dynamic assessment* (pp. 82–113). New York: Guilford.

Carroll, J. (1993). *Human cognitive abilities.* Cambridge: Cambridge University Press.

Chaiklin, S. (2003). The Zone of Proximal Development in Vygotsky's analysis of learning and instruction. In A. Kozulin, B. Gindis, V. Ageyev, and S. Miller (Eds.), *Vygotsky's educational theory in cultural context* (pp. 39–64). New York: Cambridge University Press.

Chazan, D. (2000). *Beyond formulas in mathematics and teaching dynamics of the high school algebra classroom.* New York: Teachers College Press.

Cioffi, G., and Carney, J. (1983). Dynamic assessment of reading disabilities. *The Reading Teacher, 36,* 764–768.

Cohen, J. (1988). *Statistical power analysis for the behavioral sciences.* Hillsdale, NJ: Lawrence Erlbaum Associates.

Coolidge, J. L. (1942). *A history of geometrical methods.* Oxford: Oxford University Press.

Crosby, A. (1997). *The measure of reality: Quantification and Western society.* New York: Cambridge University Press.

Cummins, J. (2000). Learning to read in a second language. In S. Shaw (Ed.), *Intercultural education in European classroom.* Stone-on-Trent, UK: Trentham Books.

Davydov, V. (1988a). Problems of developmental teaching: Part 1. *Soviet Education, 8,* 15–97.

Davydov, V. (1988b). Problems of developmental teaching: Part 2. *Soviet Education, 9,* 3–83.

Davydov, V. (1988c). Problems of developmental teaching: Part 3. *Soviet Education, 10,* 3–77.

Davydov, V. (1990). *Types of generalization in instruction.* Reston, VA: National Council of Teachers of Mathematics.

Davydov, V. (1992). The psychological analysis of multiplication procedures. *Focus on Learning Problems in Mathematics, 14,* 3–67.

Davydov, V., Gorbov, S., Mikulina, G., and Saveleva, O. (1999). *Mathematics: Class 1.* Binghamton, NY: State University of New York.

Davydov, V., and Tsvetkovich, Z. (1991). On the objective origin of the concept of fractions. *Focus on Learning Problems in Mathematics, 14,* 3–67.

Descartes, R. (2001). *Discourse on method, optics, geometry, and meteorology.* P. J. Oscamp (Tr.). Indianapolis, IN: Hachett.

Duckworth, E. (1987). *The having of wonderful ideas.* New York: Teachers College Press.

Ekstrom, R. B., French, J. W., Harman, H. H., and Dermen, D. (1976). *Manual for kit of factor-referenced cognitive tests.* Princeton, NJ: Educational Testing Service.

Ekstrom, R. B., French, J. W., and Harman, H. H. (1979). Cognitive factors: Their identification and replication. *Multivariate Behavioral Research Monograph, 79*(2), 1–85.

Eves, H. (1990). *An introduction to the history of mathematics.* Austin, TX: Holt, Rinehart & Winston.

Feuerstein, R. (1990). The theory of structural cognitive modifiability. In B. Presseisen (Ed.), *Learning and thinking styles* (pp. 68–134). Washington, DC: National Education Association.

Feuerstein, R., Krasilovsky, D., and Rand, Y. (1978). Modifiability during adolescence. In J. Anthony (Ed.), *The child and his family: Children and their parents in a changing world* (pp. 197–217). London: Wiley.

Feuerstein, R., Rand, Y., and Hoffman, M. (1979). *Dynamic assessment of retarded performers.* Baltimore, MD: University Park Press.

Feuerstein, R., Rand, Y., Hoffman, M., and Miller, R. (1980). *Instrumental Enrichment.* Baltimore, MD: University Park Press.

Feuerstein, R., Rand, Y., Falik, L., and Feuerstein R. S. (2002). *Dynamic assessment of cognitive modifiability.* Jerusalem: ICELP Press.

Foshay, A. (2007). *Brief history of IEA.* International Association for the Evaluation of Educational Achievement. Retrieved July 14, 2007, from http://www.iea.nl/brief_history_of_iea.html

Fosnot, C. T. (Ed.). (1996). *Constructivism: Theory, perspectives, and practice.* New York: Teachers College Press.

Gerber, M. (2000). Dynamic assessment for students with learning disabilities. In C. Lidz and J. Elliott (Eds.), *Dynamic assessment: Prevailing models and applications* (pp. 263–292). New York: Elsevier Science.

Gillings, R. J. (1972). *Mathematics in the time of the Pharaohs.* Mineola, NY: Dover.

Ginsburg, A., Cooke, G., Leinwand, S., Noell, J., and Pollock, E. (2005). *Reassessing U.S. international mathematics performance: New findings from the 2003 TIMSS and PISA.* Washington, DC: American Institute for Research.

Glasersfeld, E. (1995). *Radical constructivism.* London: Falmer.

Guthke, J., and Beckmann, J. (2000). Learning test concept and dynamic assessment. In A. Kozulin and Y. Rand (Eds.), *Experience of mediated learning* (pp. 175–190). Oxford: Pergamon.

Guthke, J., and Wingenfeld, S. (1992). The learning test concept: Origins, state of art, and trends. In C. Haywood and D. Tzuriel (Eds.), *Interactive assessment* (pp. 64–93). New York: Springer.

Harpaz, Y. (2007). Approaches to teaching thinking: Toward a conceptual mapping of the field. *Teachers College Record, 109*(8), 1845–1874.

Hatano, G. (1997). Learning arithmetic with an abacus. In T. Nunes and P. Bryant (Eds.), *Learning and teaching mathematics: An international perspective* (pp. 209–232). Hove, UK: Psychology Press.

Haycock, K. (2001). Closing the achievement gap. *Educational Leadership, 58*(6), 6–11.

Haywood, C., and Lidz, C. (2007). *Dynamic assessment in practice.* New York: Cambridge University Press.

Henningsen, M., and Stein, M. K. (1997). Mathematical tasks and student cognition: Classroom-based factors that support and inhibit high-level mathematical thinking and reasoning. *Journal for Research in Mathematics Education, 28*(5), 524–549.

Herrenstein, R., and Murrey, C. (1994). *The bell curve.* New York: Free Press.

Hiebert, J., and Carpenter, T. P. (1992). Learning and teaching with understanding. In D. A. Grouws (Ed.), *Handbook of research on mathematics teaching and learning.* New York: Macmillan.

Hiebert, J., Stigler, J., Jacobs, J., Givvin, K., Garnier, H., Smith, M., Hollingworth, H., Monaster, A., Wearne, D., and Gallimore, R. (2005). Mathematics teaching in the United States today and tomorrow. Results from the TIMSS 1999 Video Study. *Educational Evaluation and Policy Analysis, 27,* 111–132.

Huang, G. (2000). Mathematics achievement by immigrant children. *Educational Policy Analysis Archives, 8*(25), 1–15.

Jacobson, D., and Kozulin, A. (2007). Dynamic assessment of proportional reasoning: Human vs. computer-based mediation. Paper presented at the 12th European Conference for Research on Learning and Instruction, Budapest.

Karpov, Y. (2003a). Vygotsky's doctrine of scientific concepts. In A. Kozulin, B. Gindis, V. Ageyev, and S. Miller (Eds.), *Vygotsky's educational theory in cultural context* (pp. 65–82). New York: Cambridge University Press.

Karpov, Y. (2003b). Development through the life-span: A neo-Vygotkian approach. In A. Kozulin, B. Gindis, V. Ageyev, and S. Miller (Eds.), *Vygotsky's educational theory in cultural context* (pp. 138–155). New York: Cambridge University Press.

Kinard, J. (2001). Cognitive, affective and academic changes in specially challenged inner city youth. In A. Kozulin, R. Feuerstein, and R. S. Feuerstein (Eds.), *Mediated learning experience in teaching and counseling* (pp. 55–64). Jerusalem: ICELP Press.

Kinard, J., and Kozulin, A. (2005). Rigorous mathematical thinking: Mediated learning and psychological tools. *Focus on Learning Problems in Mathematics, 22,* 1–29.

Kline, M. (1972). *Mathematical thought from ancient to modern times.* New York: Oxford University Press.

Kozulin, A. (1995). The learning process: Vygotsky's theory in the mirror of its interpretations. *School Psychology International, 16,* 117–129.

Kozulin, A. (1998a). *Psychological tools: A sociocultural approach to education.* Cambridge, MA: Harvard University Press.

Kozulin, A. (1998b). Profiles of immigrant students' cognitive performance on Raven's progressive matrices. *Perceptual and Motor Skills, 87,* 1311–1314.

Kozulin, A. (2000). Diversity of Instrumental Enrichment applications. In A. Kozulin and Y. Rand (Eds.), *Experience of mediated learning* (pp. 257–273). Oxford: Elsevier Scientific.

Kozulin, A. (2005a). *Integration of culturally different students in mainstream classes.* Paper presented at the International Conference on Inclusive and Cognitive Learning in Schools, Prague.

Kozulin, A. (2005b). *Who needs metacognition more: Teachers or students*? Paper presented at the Annual Meeting of the American Educational Research Association, Montreal.

Kozulin, A., and Garb, E. (2002). Dynamic assessment of EFL text comprehension. *School Psychology International, 23,* 112–127.

Kozulin, A., and Garb, E. (2004). Dynamic assessment of literacy: English as a third language. *European Journal of Psychology of Education, 19,* 65–78.

Kozulin, A., Gindis, B., Ageyev, V., and Miller, S. (Eds.). (2003). *Vygotsky's educational theory in cultural context.* New York: Cambridge University Press.

Kozulin, A., Kaufman, R., and Lurie, L. (1997). Evaluation of the cognitive intervention with immigrant students from Ethiopia. In A. Kozulin (Ed.), *The ontogeny of cognitive modifiability.* Jerusalem: ICELP Press.

Levi-Strauss, C. (1969). *The elementary structures of kinship.* London: Eyre.

Lidz, C., and Gindis, B. (2003). Dynamic assessment of the evolving cognitive functions in children. In A. Kozulin, B. Gindis, V. Ageyev, and S. Miller (Eds.), *Vygotsky's educational theory in cultural context* (pp. 99–118). New York: Cambridge University Press.

Luria, A. (1976). *Cognitive development: Its cultural and social foundations.* Cambridge, MA: Harvard University Press.

Lindquist, M. (Ed.). (1989). *Results from the 4th mathematics assessment of the National Assessment of Mathematical Progress.* Reston, VA: NCTM.

Ma, L. (1999). *Knowing and teaching elementary mathematics.* Mahwah, NJ: Lawrence Erlbaum.

Marzano, R. J. (2000). *Transforming classroom grading.* Alexandria, VA: Association for Supervision and Curriculum Development.

Marzano, R. J. (2006). *Classroom assessment and grading that work.* Alexandria, VA: Association for Supervision and Curriculum Development.

Mead, G. H. (1974). *Mind, self and society.* Chicago: University of Chicago Press.

Menninger, K. (1969). *Number words and number symbols.* Cambridge, MA: MIT Press.

Morris, A. (2000). A teaching experiment: Introducing 4th graders to fractions from the viewpoint of measuring quantities using Davydov's mathematics curriculum. *Focus on Learning Problems in Mathematics, 22,* 32–84.

National Commission on Mathematics and Science Teaching for the 21st Century and submitted to former U.S. Secretary of Education, Richard W. Riley. (2000). *Before it's too late.* Retrieved January 3, 2004, from http://www.ed.gov/inits/Math/glenn/report.pdf

National Council of Teachers of Mathematics. (1989). *Curriculum and evaluation standards for school mathematics.* Reston, VA: NCTM.

National Council of Teachers of Mathematics. (1991). *Professional standards for teaching mathematics.* Reston, VA: NCTM.

National Council of Teachers of Mathematics. (1995). *Assessment standards for school mathematics.* Reston, VA: NCTM.

National Council of Teachers of Mathematics. (2000). *Principles and standards for school mathematics.* Reston, VA: NCTM.

National Council of Teachers of Mathematics. (2001). *Navigating in algebra in grades 6–8.* Reston, VA: NCTM

National Research Council. (1996). *National science education standards.* Washington, DC: National Academy Press.

National Science Foundation. (2000). *America's investment in the future.* Arlington, VA: NSF.

Newton, X. (2007). Reflections on math reforms in the U.S.: A cross-national perspective. *Phi Delta Kappan, 88*(9), 681–685.

Nunes, T. (1999). Mathematics learning and the socialization of the mind. *Mind, Culture, and Activity, 6,* 33–52.

O'Donnell, B., and Taylor, A. (2006/2007). A lesson plan as professional development?: You've got to be kidding! *Teaching Children Mathematics: National Council of Teachers of Mathematics, 13*(5), 272–278.

Olson, D. (1994). *The world on paper: The conceptual and cognitive implications of writing and reading.* New York: Cambridge University Press.

Olson, S. (1999). Candid camera [Electronic version]. *Teacher Magazine,* May/June. Retrieved January 3, 2005, from www.ed.gov/inits/Math/glenn/report.pdf

Otis, A. S., and Lennon, R. T. (1996). *Otis-Lennon school ability test.* San Antonio, TX: Harcourt Brace.

Perkins, D., and Salomon, G. (1989). Are cognitive skills content bound? *Educational Researcher, 18,* 16–25.

Piaget, J. (1947/1969). *Psychology of intelligence.* Totowa, NJ: Littlefield, Adams & Co.

Piaget, J. (1970). *Structuralism.* New York: Basic Books.

Pullan, J. M. (1968). *The history of abacus.* London: Books That Matter.

Reese, C., Miller, K., Mazzeo, J., and Dossey, J. (1997). *NAEP 1996 mathematics report card for the nation and the states.* Washington, DC: National Center for Education Statistics.

Reys, B., and Lappan, G. (2007). Consensus or confusion?: The intended math curriculum in state-level standards. *Phi Delta Kappan, 88*(9), 676–680.

Rogoff, B. (1990). *Apprenticeship in thinking.* Oxford: Oxford University Press.

Rogoff, B. (2003). *The cultural nature of human development.* New York: Oxford University Press.

Rothman, R. (2006). (In)Formative assessments: New tests and activities can help teachers guide student learning. *Harvard Education Letter, 22*(6). Retrieved January 3, 2006, from http://www.timeoutfromtesting.org/0519_article_harvard.php

Schmittau, J. (2003). Cultural historical theory and mathematics education. In A. Kozulin, B. Gindis, V. Ageyev, and S. Miller (Eds.), *Vygotsky's educational theory in cultural context* (pp. 225–245). New York: Cambridge University Press.

Schmittau, J. (2004). Vygotskian theory and mathematical education. *European Journal of Psychology of Education, 19*, 19–44.

Schoultz, J., Saljo, R., and Wyndhamn, J. (2001). Heavenly talk: Discourse, artifacts and children's understanding of elementary astronomy. *Human Development, 44*, 103–118.

Scribner, S., and Cole, M. (1981). *Psychology of literacy.* Cambridge, MA: Harvard University Press.

Sertima, I. V. (Ed.). (1984). *Blacks in science, ancient and modern.* New Brunswick, NJ: Transaction Books.

Sfard, A. (1998). On two metaphors for learning and the danger of choosing just one. *Educational Researcher, 27*(2), 4–13.

Shulman, L. S. (1986). Those who understand: Knowledge growth in teaching. *Educational Researcher, 15*(1), 4–14.

Smeltzer, D. (2003). *Man and number.* Mineola, NY: Dover.

Smith, D. E. (1958). *History of mathematics.* New York: Dover.

Smith, I., Martin, M., Mullis, I., and Kelly, D. (2000). *Profiles of student achievement in science at the TIMSS international benchmarks.* Chestnut Hill, MA: Boston College.

Stafylidou, S., and Vosniadou, S. (2004). The development of students' understanding of fractions. *Learning and Instruction, 14*(5), 503–518.

Stech, S. (2007). Special features of cognition and learning in the school context. *Transylvanian Journal of Psychology, 2*, 63–76.

Sternberg, R. (1980). Sketch of a componental subtheory of human intelligence. *Behavioral and Brain Sciences, 3*, 573–584.

Sternberg, R., and Grigorenko, E. (2002). *Dynamic testing.* New York: Cambridge University Press.

Stigler, J., Chalip, L., and Miller, K. F. (1986). Consequences of skill: The case of abacus training in Taiwan. *American Journal of Education, 94*(4), 447–479.

Stigler, J., and Hiebert, J. (1997). Understanding and improving classroom mathematics instruction: An overview of TIMSS video study. *Phi Delta Kappan, 79*(1), 14–21.

Sunderman, G., and Orfield, G. (2006). Domesticating a revolution: No Child Left Behind reforms and state administrative response. *Harvard Educational Review, 76*, 526–555.

Vamvakoussi, X., and Vosniadou, S. (2004). Understanding the structure of the set of rational numbers. *Learning and Instruction, 14*(5), 453–467.

Verschaffel, L. (1999). Realistic mathematical modeling and problem solving in the upper elementary school. In J. H. M. Hamers, J. E. H. Van Luit, and B. Csabo (Eds.), *Teaching and learning thinking skills* (pp. 215–240). Lisse: Swets & Zeitlinger.

Vinovskis, M. A. (2007). *Overseeing the nation's report card: The creation and evolution of the National Assessment Governing Board (NAGB).* Washington, DC: National Assessment Governing Board, U.S. Department of Education.

Vygotsky, L. (1978). *Mind in society.* Cambridge, MA: Harvard University Press.

Vygotsky, L. (1979). Instrumental method in psychology. In J. Wertsch (Ed.), *The concept of activity in Soviet psychology* (pp. 134–143). Armonk, NY: Sharpe.

Vygotsky, L. (1986). *Thought and language* (rev. ed.). Cambridge, MA: MIT Press.

Vygotsky, L. (1998). *Child psychology: The collected works of L. S. Vygotsky* (Vol. 5). New York: Plenum.

Vygotsky, L., and Luria, A. (1993). *Studies on the history of behavior.* Hillsdale, NJ: Lawrence Erlbaum.

Werner, H., and Kaplan, B. (1984). *Symbol formation.* Hillsdale, NJ: Lawrence Erlbaum.

Wixson, K. K., Dutro, E., and Athan, R. G. (2003). The challenge of developing content standards. *Review of Research in Education, American Educational Research Association, 27,* 69–82.

Wood, D. (1999). Teaching the young child: Some relationships between social interaction, language, and thought. In P. Lloyd and C. Fernyhough (Eds.), *Lev Vygotsky: Critical assessments* (Vol. 3, pp. 259–275). London: Routledge.

Xie, X. (2002). *The cultivation of problem-solving and reason in NCTM and Chinese national standards.* Nanjing: Nanjing Normal University, School of Education.

Zaslavsky, C. (1984). The Yoruba number system. In I. V. Sertima (Ed.), *Blacks in science, ancient and modern* (pp. 110–126). New Brunswick, NJ: Transaction Books.

Zuckerman, G. (2003). The learning activity in the first years of schooling. In A. Kozulin, B. Gindis, V. Ageyev, and S. Miller (Eds.), *Vygotsky's educational theory in cultural context* (pp. 177–199). New York: Cambridge University Press.

Zuckerman, G. (2004). Development of reflection through learning activity. *European Journal of Psychology of Education, 19,* 9–18.

Index

www.ingramcontent.com/pod-product-compliance
Ingram Content Group UK Ltd.
Pitfield, Milton Keynes, MK11 3LW, UK
UKHW040811180125
453697UK00004B/21

9 780521 876858